Atomic–Molecular Ionization by Electron Scattering

For analyzing the physical aspects of electron collisions with atoms and molecules, an understanding of cross sections, especially of the ionization cross sections, for the impacting electrons is necessary. Providing a detailed and up-to-date discussion to the theory and applications of electron scattering, this text equips readers with the basic concepts of quantum scattering theories, including complex scattering potential method, cross section results for atoms including those of inert gases, atomic oxygen, nitrogen, and carbon, and also for the less explored atomic systems.

Study of electron scattering phenomena for diatomic and common molecules, polyatomic molecules, and radicals including hydrocarbons, fluorocarbons, and other larger molecules, together with relevant radical species, is discussed in detail. This book discusses applications of electron impact ionization and excitation in gaseous or plasma and condensed matter. Recent advances in the field of electron molecule scattering and ionization for polyatomic molecules are covered extensively.

K. N. Joshipura is former Professor and Head, Department of Physics, Sardar Patel University, Vallabh Vidyanagar, India. He has more than 40 years of teaching and research experience and has published more than 130 papers in national and international journals. His areas of research include theoretical studies on electron scattering by atomic–molecular targets in general and metastable and radical species in particular.

Nigel Mason is Professor and Head, School of Physical Sciences, University of Kent, UK. He has published more than 350 papers in journals of international repute. His current research includes experimental studies of molecular formation on dust grains in the interstellar medium and on planetary surfaces as well as electron induced processes in outer-space environments.

Atomic–Molecular Ionization by Electron Scattering

Theory and Applications

K. N. Joshipura
Nigel Mason

CAMBRIDGE
UNIVERSITY PRESS

CAMBRIDGE
UNIVERSITY PRESS

University Printing House, Cambridge CB2 8BS, United Kingdom

One Liberty Plaza, 20th Floor, New York, NY 10006, USA

477 Williamstown Road, Port Melbourne, vic 3207, Australia

314 to 321, 3rd Floor, Plot No.3, Splendor Forum, Jasola District Centre, New Delhi 110025, India

79 Anson Road, #06–04/06, Singapore 079906

Cambridge University Press is part of the University of Cambridge.

It furthers the University's mission by disseminating knowledge in the pursuit of education, learning and research at the highest international levels of excellence.

www.cambridge.org
Information on this title: www.cambridge.org/9781108498906

First published 2019

Printed in India by Rajkamal Electric Press

A catalogue record for this publication is available from the British Library

ISBN 978-1-108-49890-6 Hardback

Additional resources for this publication at www.cambridge.org/9781108498906

To our teachers and students ...

CONTENTS

FIGURES

TABLES

FOREWORD I

Electron collisions with atoms and molecules are commonplace. In the natural world they occur in lightning strikes, aurorae, and the Earth's ionosphere in general; outside our planet they are important for similar processes in other planets. The glow of Jupiter's aurora can clearly be seen using telescopes from the Earth. Electron collisions also form a primary process in cometary tails that are bathed in the solar wind, and in many other astrophysical processes. Plasma is the fourth state of matter which involves partial ionization of the atomic and molecular components. Plasmas occur naturally in flames, stars, and elsewhere. Humankind has increasingly harnessed the power of electron collisions in many ways: to start cars with spark plugs, in the traditional light bulb, and in many lasers. Much of modern industry is driven by the use of electron collisions to create plasmas which etch silicon and other materials into ever more complex structures or to provide surface coatings to alter, enhance, or protect the properties of materials. The quest to harness the Sun's power on Earth via fusion involves making a vast hot plasma with a wealth of electron collision processes requiring detailed study. In the current century it has also been realized that the damage experienced by bio-systems as a consequence of all types of high energy particles and radiation is predominantly caused by collisions involving secondary electrons. These electrons are created by the ionizing effect of the original high-energy collision particle independent of the nature of the colliding species. In medical applications these collisions can be harmful, causing double strand breaks of DNA, or beneficial as in radiation therapy, which is widely used to exorcize malignant tumours.

Electrons colliding with atoms and particularly molecules can initiate a variety of processes. Probably the most important of these is the creation of ions (charged species) either through impact ionization or by electron attachment leading to positively and negatively charged ions respectively. These ionized species are chemically active and act as initiators of many of the processes mentioned above.

Understanding, controlling, and utilizing the full power of the collisions of electrons with atoms and molecules has thus become a major objective of many physical scientists but with benefits and applications well beyond the field of physical science. However, electron collisions with atoms and molecules are governed by the laws of quantum mechanics. This means that the collisions do not obey the everyday laws of collisions such as those found on the billiard table but instead behave in subtle ways which require understanding of the physics at the atomic level. It is just over a hundred years since Franck and Hertz famously demonstrated that the energy levels of the mercury atom were quantized (i.e. discrete), by studying the effects of electron collisions. Since then our understanding of such processes and our ability to control and harness them has made huge progress. This progress has been achieved through the combined application of greatly improved measurement techniques and enhanced theoretical methods, fortified by the power of modern computers. Of course electron collisions also provide fundamental insights into the nature of the collision partner. There is thus now a wealth of knowledge on the processes that follow electron collisions with atoms and molecules as captured both in data sets that can form the input to detailed models of electron collisions and through general principles of what drives these processes.

I therefore welcome this book. It is written by two physicists who have studied electron collision physics over many decades, gaining important insights into ionization and other processes through applications of novel theoretical and experimental methods. The book places their years of experience in a single volume in an easy and accessible manner for the education and enjoyment of the reader.

Jonathan Tennyson

FRS
Massey Professor of Physics, University College London
London (UK)

FOREWORD II

To study any system, an interaction with the system is essential. For microscopic objects and systems, which cannot be seen by the naked eye, usually the electron (or photon) beam, with known characteristics, acts as the probe. At a micro-distance, the projectile particles interact/collide with the object and subsequently they are scattered in all possible directions. The scattered particles, electrons in our case, carry the signature of interaction with the object (target). Hence, the measurement of the differential cross sections, $I(\theta, \emptyset)$ over all possible directions, yields the total collisional cross section $\sigma(Ei)$ as a function of the incident energy Ei of the electrons. For the atomic targets, the collisions are elastic as well as inelastic, including ionization. For the molecules, the additional processes like dissociation and the dissociative ionization, etc., are also possible. Besides, the dissociative components may be in the excited state. Myriad phenomena arising out of electron scattering make the study very interesting from the manifold view-points of theory, experiment, as well as applications.

In the present book, written by Professors K. N. Joshipura and Nigel Mason, collisions of electrons with atoms are briefly discussed initially. In the greater part of the book, molecules are considered as targets. The study is extended to the radicals and the metastable molecules, while a few molecules of biological interest are also considered. The applications of various scattering cross sections to diverse fields like astrophysics, astrochemistry, nanotechnology, etc., are described in the last chapter.

Both the authors are experienced players in the field of electron–atom–molecule collisions. They have a good number of publications to their credit. From the theory as well as application point of view the present book should be quite useful.

S. P. Khare

Former Pro Vice Chancellor and Emeritus Professor
C. C. S. University, Meerut (India)

PREFACE

Electrons are ubiquitous in nature and throughout modern industry, and therefore there are varieties of situations in which electrons interact with atoms and molecules producing diverse physical and chemical phenomena. Extensive studies, both experimental and theoretical, have been carried out on the interactions of electrons with different atomic and molecular targets; indeed the last few decades have witnessed rapid developments in the techniques and methodology for exploring electron–atom/molecule scattering. The wider recognition of the role of fundamental electron interactions in natural phenomena (for example, the observation of aurorae on other planets and the contribution of electron interactions in astrochemistry), in underpinning novel technologies such as Focussed Electron Beam Induced Deposition (FEBID), and as a major source of radiation damage by ionizing radiation has led to an increase in the size of the international community studying electron collisions in all phases of matter.

In this book, our aim is to provide an overview of the field with a focus on theoretical methods used to describe the collisions of intermediate to high energy (exceeding about 15 eV) electrons. The book has six chapters and begins with a discussion of the subject by outlining the necessary textbook background on atoms, molecules, and quantum scattering theories. Attention has been devoted (in Chapter 1) to atomic sizes or 'radii' – something that is normally missing in most books and reviews of this kind. A brief survey of atomic radii, running across the periodic table of elements, is outlined.

The major part of this monograph provides an up-to-date review of electron scattering from atoms and molecules, summarizing recent publications. Although the title of the present book mentions ionization specifically, the contents are comprehensive in that we highlight several important inelastic processes ocurring in the background of elastic scattering. For many atoms and a large number of molecules, recent theoretical results are discussed along with experimental and other data, and wherever possible recommended data are presented to provide the user with data sets for models and simulations of processes in which electron interactions play a significant role.

The book also provides a summary of basic and most used electron scattering theories and, in particular, discusses an approximate theoretical formalism, called 'Complex Scattering Potential – ionization contribution' method, developed by the authors to derive ionization cross sections for a large variety of atomic and molecular systems typically from the first ionization energy onwards, up to 2000 eV or so.

Results for common/atmospheric atoms and molecules are presented in Chapters 2 to 4, while results for polyatomic and/or exotic molecules, including hydrocarbons, fluorocarbons, and biomolecules, etc., are presented in Chapter 5. Of particular interest are the reactive radicals and long-lived metastable species for which experiments are scarce or non-existent and hence the theoretical cross section data hold more significance to the user community. Attempts have also been made, separately for atoms and molecules, to correlate the dynamic quantities, i.e., total cross sections, with some of the static properties of a variety of atomic and molecular systems.

The final chapter, Chapter 6, discusses many fields of science and technology where electron interactions with atoms and molecules play a prominant role. Nature provides large veritable laboratories in the form of the atmospheres of Earth, Mars, Venus, Jupiter, Saturn, etc., and their satellites to explore the role of electron interactions with atoms and molecules whilst their importance has also been revealed recently in comets and other astronomical systems, including the atmospheres of exoplanets.

Electron scattering is a dominant process in many technological fields such as gaseous electronics and electrical discharges, mass spectrometry, lasers and plasma systems, etc. Indeed many of the plasma and related nano-fabrication techniques such as the emerging FEBID and EUV lithography are governed by electron induced processing and scattering. We briefly discuss the role of electron scattering in regulating plasma confinement in fusion plasmas. We also briefly discuss the interaction of electrons with larger biomolecular systems, since it is now recognized that secondary electron emission, and subsequent electron regulated damage to cellular DNA, is a determining factor in radiation damage and, if controlled, may provide new treatment processes in clinical radiotherapy.

For completeness we also discuss electron scattering in the condensed phase of matter and consider scattering by the electron anti-particle, the positron. Scattering of positrons with several atomic and molecular targets is reviewed, mainly in the spirit of providing a comparison with electron scattering.

Thus, in essence, our plan in this work is to place before the scientific community an updated overview of the status of electron interactions with atoms and molecules and the current theoretical methods for exploring such effects. Whilst we emphasize high energy theoretical research, we provide the reader with a comprehensive set of references from which they can explore the field further. We have also demonstrated the wide range of applications of electron scattering from atoms and molecules and hope the data compilation will be of use to these communities whilst providing them with a description of the underlying physics. We recognize that we may have missed some results and as new data are being published all the time this book will need updating. Suggestions and comments are most welcome in this regard.

June 2018 Kamalnayan N. Joshipura

Nigel Mason

ACKNOWLEDGMENTS

The authors are highly thankful to Professor J. Tennyson (University College London, UK) and Professor S. P. Khare (Retd., CCS University, Meerut, India) for kindly writing the Forewords to this book. Our special thanks are due to Dr Karl K. Irikura (NIST, USA) for sharing his unpublished data.

K. N. Joshipura: I take this opportunity to thank my teachers Professors H. S. Desai and M. P. Maru. It is a pleasure to thank my students Drs Minaxi Vinodkumar, Bobby Antony, Chetan Limbachiya – who are now independent researchers on their own – together with Drs Foram Joshi, Siddharth Pandya, Umang Patel, Bhushit Vaishnav, Mohit Swadia, and Harshit Kothari for rendering help in the preparation of this book. My thanks are also due to Professors P. C. Vinodkumar, K. L. Baluja, and M. Khakoo. I also thank my wife, Jagruti, for her care and patience and her active support in all of my academic pursuits.

Nigel Mason: I wish to thank my many colleagues with whom I have collaborated in the field of electron–atom and electron–molecule collisons and their applications. Particular thanks to my 'mentors' and then colleagues Professors M. Allan, E. Illenberger, T. D. Maerk, L. Sanche, E. Krishnakumar, H. Hotop, F. Gianturco, D. Field, and the sadly deceased J. Skalny, giants of the field but also inspiring people. I must also thank my many students and research fellows – who really do the work! It has been a pleasure to see them develop as independent researchers, many of whom are now leading the next generation of electron–atom/molecule studies, particularly Drs S. Eden, M. A. Smialek-Telega, Professor P. Vieira, and my Indian 'family', Minaxi Vinodkumar, Bobby Antony, Chetan Limbachiya, and Bhala Sivaraman. Finally, my thanks to my wife, Jane, who has borne my passion for science and many absences with patience and understanding; I would not have done so much without her support.

We thank the authors/publishers of other sources for permission to reproduce their figures in this book, meant for educational and academic purposes. Omission if any in this regard, though inadvertent, may please be pointed out to the present authors/publisher.

ATOMS AND MOLECULES AS BOUND QUANTUM SYSTEMS

This book will examine and explain the scattering processes of intermediate to high energy electrons in collisions with a variety of atoms and molecules. It will cover the impact energy range from the first ionization energy of most atoms and molecules (about 10–15 eV) to energies at or above 2000 eV. Since the response of an atomic or molecular target to the incident electron depends on both its impact energy and the structural properties of the target itself, it is appropriate to begin with an overview of atoms and molecules as bound quantum systems. Historical references will therefore be pertinent in this context.

1.1 INTRODUCTION – A BRIEF HISTORY OF THE STUDY OF ELECTRON SCATTERING

The beginning of the twenty-first century should be considered as a centenary marker for a number of fundamental developments in physics. The year 2014 was the centenary of the first announcement of the famous Franck and Hertz Experiment. Conducted in 1914, the Franck–Hertz Experiment is widely regarded as the one that provided validation for the Bohr theory of atomic structure, itself published only a year back, in 1913. It should also be viewed as the first quantitative experiment in electron scattering and the birth of scientific study of atomic and molecular phenomena by collisions. The original Franck–Hertz experiment passed a beam of electrons through mercury vapour with subsequent detection of the kinetic energy of the transmitted electrons. A characteristic series of peaks were observed, separated by 4.9 eV (254 nm), precisely the energy needed to achieve the first excited state of the mercury atom (Figure 1.1). Thus an incident electron, scattered by the atom, transferred its kinetic energy to the target, resulting in exciting the atom. Measuring the energy lost by scattering electrons makes it possible to determine excitation energies in the target, and this is the basis of the now standard analytical technique of Electron Energy Loss Spectroscopy (EELS).

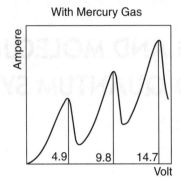

Figure 1.1 The typical Franck–Hertz electron curve showing the presence of a series of peaks separated by 4.9 eV (254 nm) characteristic of the excitation energy of the first electronic state of mercury

Once the true nature of the Franck–Hertz experiment was understood (the authors originally believing it to be an ionization phenomenon), a series of electron transmission experiments were performed. Ramsauer (1921) developed a more sophisticated apparatus using magnetic fields to collimate a beam of electrons whose energy could be more accurately defined (Figure 1.2). At low energies (less than 1 eV) when electrons passed through a chamber containing one of the heavier rare gases (Argon, Krypton, and Xenon) the gas appeared to be 'transparent' with little or no scattering occurring, even for elastic scattering (Figure 1.2b). Such a phenomenon could not be explained by any classical theory which, from its 'billiard balls approach', would lead to expectations of both significant scattering and, at lower energies, with correspondingly longer interactions times, a larger scattering cross section. However, by applying the Schrodinger equation of quantum mechanics and considering the electron projectile as an incident wave composed of partial waves characterized by the momentum of the electron, this Ramsauer–Townsend minimum (Townsend conducted similar contemporary experiments at Oxford in the UK) could be explained. Accordingly, quantum mechanics could be employed to study electron–atom/molecular interactions and calculate scattering cross sections which could be validated by experiment.

Thus it became possible to quantify electron collisions and use these cross sections to model natural and laboratory phenomena. Mott and Massey, in the UK, are considered pioneers in ushering in the era of theoretical collision physics (Mott and Massey 1965). The first applications of this theory coincided with the technological development of radar and the discovery of the ionosphere. Ionospheric models and a physical description of that most magnificent of natural phenomena, the aurora, were based on theoretical descriptions and experimental measurement of electron–atom and electron–molecule collisions leading to excitation and ionization of the target species.

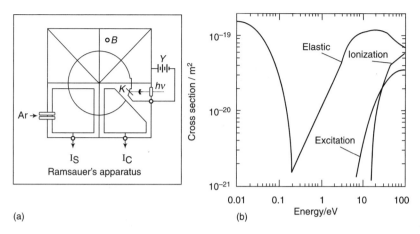

Figure 1.2　(a) The Ramsauer apparatus used to measure low energy scattering from atoms and molecules, revealing (b) the Ramsauer–Townsend minimum in the total and elastic scattering cross section of the rare gases

A major breakthrough came in the 1960s with the development of 'monochromated' electron beams, typically beams with resolution of less than 100 meV (milli electron volts), capable of resolving vibrational structure in energy loss spectra. Such high resolution electron beams also revealed unexpectedly fine structures in collision cross sections. For example, a sharp dip in the elastic scattering cross section of electrons from helium at 19.317 eV (Figure 1.3). Such a phenomenon, below the first excited state of helium (19.819 eV), appeared to have no explanation but similar effects were seen in other atoms and molecules. Schulz (1963, 1973) proposed that these 'resonances' could be explained if the electron was temporarily 'captured' by the target to form a 'temporary negative ion (TNI)' with a life of only femtoseconds or picoseconds before decaying by auto-detachment, thereby displaying Delayed Elastic Scattering, i.e.,

$$e + AB \rightarrow AB^- \rightarrow AB + e$$

An alternative route of decay leads to the fragmentation of the TNI to form an anionic product

$$AB^- \rightarrow A + B^-$$

a process known as Dissociative Electron Attachment or DEA. DEA cross sections are strongest for targets containing halogen atoms (since they are electronegative) but may be large in compounds containing 'pseudo-halogen' chemical groups such as –CN. DEA occurs naturally in the Earth's ionosphere where O^- anions are formed by DEA to molecular oxygen.

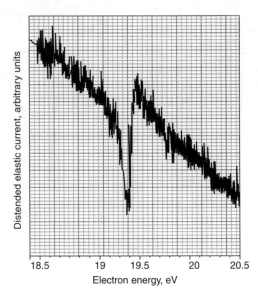

Figure 1.3 The first observation of an electron scattering resonance in helium (Schulz 1963)

The desire to understand the dynamics of electron scattering led to ever more sophisticated experiments, including the study of electron–photon coincidences (Kleinpoppen et al. 2013) to probe electron induced excitation (mainly of rare gases); electron–ion coincidence experiments (Maerk and Dunn 2010) to explore electron induced ionization and possible correlation between incident, scattered, and ejected electrons; the study of spin polarized electrons scattering from atoms and molecules (Gay 2009); and electron scattering in the presence of laser fields (Mason 1993); to couple electron scattering with photon excitation of the target. This included the study of 'electron induced' de-excitation or 'super-elastic' scattering and simultaneous electron–photon excitation. Such detailed experiments provided challenging tests of theoretical concepts about electron collisions, and required commensurate development of theory. Noteworthy progress accrued from coupling quantum mechanics based depictions of the scattering, with quantum chemistry based depictions of the target structures. Several 'formalisms' were developed and have stood the test of time with the R-matrix techniques (Burke 2011) derived from nuclear physics perhaps being the most well-known. However, there are many other methods with individual strengths and weaknesses and applicability to one scattering process or another. Carsky and Curik (2011) may be referred to for comprehensive information.

By the 1990s it was recognized that electron collisions were responsible for many diverse phenomena and thus the numbers of atomic and molecular targets to be investigated grew exponentially. For example, with the discovery of the 'ozone hole' and growth of awareness on global warming, certain molecules were identified as having 'global warming potential' and 'ozone depletion potential'. It was in this context that electron collision data was required for many molecular species that had not previously been studied, such as ozone, and chloro- and fluorocarbons. The recognition that electrons play a crucial role in radiation damage to biological systems has led to electron studies with bio-macromolecules including DNA. Bodies in the Solar System such as the Earth, other planets and their satellites, comets, etc., are nature's wonderful laboratories where electron–atom/molecule collisions take place continuously. With the launching of various space and planetary missions, electron collision studies have found fresh relevance. The recent Cassini-Huygens mission to Saturn and its moons has revealed that Titan (considered similar to early primordial Earth) has rich ionospheric chemistry. Electron induced processes are important in the upper atmospheres of Jupiter and Saturn where aurora have been observed and there is recent interest in the atmosphere of Mars, where electron collision based ionization is known to play a role in airglow as observed by orbiting spacecraft. Similarly, the recent Rosetta mission to a comet has revealed that electron collisions play a major role in many physical and chemical processes prevalent on these intriguing celestial bodies.

Electron scattering studies are also providing useful inputs in technological fields such as gaseous electronics and electrical discharges, mass spectrometry, lasers and plasma systems, to name just a few. A technique worthy of mention is known as FEBID (Focused Electron Beam Induced Deposition) where a highly collimated beam of very high energy electrons is directed along with a feed gas onto the target substrate. This results in the formation of metallic nanostructures on the surface that are essential for latest generation electronic devices. Electron collisions are also at the core of many plasma tools including those now used in surgery and dentistry, and play an important role in plasmas that provide lighting.

It is therefore amazing to see how the poineering experiment of Franck and Hertz, just over a hundred years ago, has culminated in the application of electron scattering from so many diverse atoms and molecules in myriad applications in science and technology. The role and applications of electron scattering constitute the final chapter of this book, but meanwhile our main focus is to examine the mechanisms of electron scattering and present a detailed overview of our current knowledge of such scattering by the many atomic and molecular targets.

This book is, therefore, a systematic compilation of the studies of inter-particle *collisions,* events of rather short duration, wherein two particles interact mutually. *Scattering,* resulting from a collision, is the deflection of the incident beam or the projectile particles caused by a target system. It is customary to use the term 'collision' or 'scattering' or 'collisional interaction' interchangeably. The same applies to discourses on electron impact phenomena or processes. Here, we examine the resulting change in the states of the scattering systems, i.e., see what happens to both the colliding electron and the target atom or molecule. As such, we follow the footsteps of many eminent scientists and their works, such as the earliest treatises, namely, *Theory of Atomic Collisions,* by Mott and Massey (1965), and well-known books written by Bransden and Joachain (2003), Joachain (1983), and Khare (2001).

At the very beginning of any study of electron scattering, it is appropriate to view an atom (or a molecule) as a quantum mechanically bound system composed of electrons and nucleus (or nuclei). A brief survey of atomic and molecular structure properties is therefore necessary in order to develop a basic understanding of the targets for electron scattering. Hydrogen and helium atoms are dealt with in some detail, while other atoms and molecules have been given briefer treatment.

1.2 THE HYDROGEN ATOM AND MULTI-ELECTRON ATOMS

A one-electron system like the hydrogen atom is an exactly solvable problem in quantum mechanics (Bransden and Joachain 2003). The corresponding energy eigenvalues and eigenfunctions are derived from the analytical solution of the time-independent Schrödinger equation,

$$\left[\frac{-\hbar^2 \nabla^2}{2m} - \frac{Ze^2}{4\pi\varepsilon_0 r} \right] \psi(r) = E\psi(r) \tag{1.1}$$

In this equation, r (with $r = |r|$) is the radial coordinate of the electron (having mass m) with reference to the proton (nucleus), ∇^2 is the Laplacian operator and Z is the atomic number. The total energy of the system is E and $\psi(r) = \psi_{n\ell m}(r, \theta, \varphi)$ is the stationary wave function of the atomic electron. The principal quantum number is represented by n, the orbital angular momentum quantum number by ℓ, and the corresponding projection quantum number by m_ℓ or simply m. These quantum numbers have the standard discrete values. (Please see also Box 1.1). In the case of the hydrogen atom, the energy eigenvalues E_n derived from equation (1.1) can be expressed by the exact formula

$$E_n = -\frac{13.6}{n^2} \text{ eV} \tag{1.2}$$

The energy level diagram of the hydrogen atom, generated from equation (1.2), is shown in Figure 1.4.

The ground state wave function of this atom written as $\psi_{100}(r)$ or ψ_{1s}, also referred to as the 1s orbital, is

$$\psi_{1s}(r) = \frac{1}{\sqrt{(\pi a_0^3)}} \exp\left(-\frac{r}{a_0}\right) \tag{1.3}$$

From equation (1.3) one finds that in this case the electron charge density is $\rho(r) = e |\psi_{1s}(r)|^2$ and this gives us the radial charge density $D(r) = 4\pi r^2 \rho(r)$, a quantity that peaks at the first Bohr radius a_0, which is a typical length on the atomic scale (Box 1.1). There are a few important observations on the H atom.

- There is a large energy difference between the ground state $n = 1$(1s) and the first excited states $n = 2$ (2s, 2p), but the energy levels get closer together as n increases.

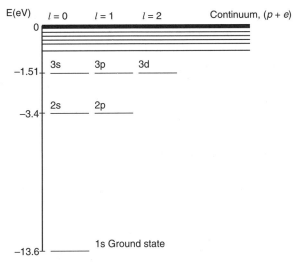

Figure 1.4 Energy levels of the Hydrogen atom, from equation (1.2)

BOX 1.1: ATOMIC UNITS (au) AND SYMBOLS

Standard symbols are used, as follows:

$\hbar = h/2\pi$, with h = Planck's constant; m_e = electron rest-mass (symbol 'm' is also used, not to be confused with the projection quantum number m); e = electron charge; ε_0 = free-space permittivity;

Å stands for the Angstrom unit of length, 1 Å = 10^{-10} m; and

a_0 = the usual first Bohr radius of the H(1s) atom = 0.529×10^{-10} m = 0.529 Å.

In the atomic-molecular regime atomic units (au in short) are defined by assuming $\hbar = m_e = e = 1$, also $4\pi\varepsilon_0 = 1$. Thus in atomic units, the Bohr radius, $a_0 = 1$ and is taken as a unit of length. In au the two units of energy adopted are the Rydberg (R) and the Hartree with 1 R= 13.6 eV, the ground-state ionization energy of atomic hydrogen, while 1 Hartree = 2 R = 27.2 eV. 1 electron Volt (eV)=1.602×10^{-19} J.

For an electron in an atom the usual quantum numbers n, ℓ, and m_ℓ have been defined in preceding text. We have the principal quantum number n = 1,2,3,4 ... corresponding to shells K, L, M, N, etc., while ℓ = 0, 1, 2, 3....(n – 1) are referred to by code letters s, p, d, f, etc. For each ℓ, the projection quantum number m_ℓ admits (2ℓ + 1) values from +ℓ to –ℓ, including zero. For an atom with number of electrons N ≥ 2, the quantum number for electronic total orbital angular momentum admits values L = 0, 1, 2, 3, ... etc., and the corresponding states are denoted by upper case letters S, P, D, F, ... etc. Thus the ground state of the two-electron atom like helium is denoted by $^1S_0$, where the left superscript '1' stands for total spin singlet state, while the right subscript '0' indicates the total angular momentum quantum number J = L + S = 0.

- The ground state (also called first) ionization energy is I_{1s} = 13.6 eV but the ionization energy is only 3.4 eV for n = 2 (Figure 1.4). There are infinitely many discrete excited states of atomic hydrogen between the first excited state and the ground-state ionization potential. The dipole selection rule governing transitions in the 1-electron atom states that, if the ℓ-value changes by 1, i.e., if $\Delta\ell = \pm 1$, then the transition is dipole allowed. This also implies the change in parity of the system wave function. Thus, the transition 1s $\leftrightarrow$ 2p is dipole allowed and the radiative life time of the 2p state is typically τ(2p) ~ 10^{-9} s. H atoms in 2p state emit the famous Lyman - α line with a wavelength of about 121.3 nm or a photon energy close to 10.2 eV. In contrast, the transition 1s $\leftrightarrow$ 2s is forbidden and hence the H-atom in 2s state is metastable with a very large radiative lifetime τ(2s) of about 1/7s. This has important consequences as the excited metastable atom can act as an energy store in the embedding medium. For transitions induced by incident electrons there are no selection rules and therefore an electron can excite the ground state of the atom to a metastable state.

- Highly excited states or high 'n' states (n ~ 100) are known as Rydberg states and they can exist in certain astrophysical systems.

- Transition probabilities from the ground state to different excited states, expressed in terms of oscillator strengths, have been discussed in advanced textbooks on this subject (e.g., Bransden and Joachain 2003). It should be noted that for atomic hydrogen the continuum oscillator strength is close to the sum of oscillator strengths of all discrete transitions, and the latter are clearly dominated by the final state n = 2.

- The continuous positive energy regime or continuum (Figure 1.4) corresponds to a system of a free proton (nucleus) and a free electron. So how do we know whether a free proton and a free electron coming closer together form a bound system, i.e., a H atom, or would they just scatter apart from one another? The answer depends on the total energy of the system. In a bound state the total energy is negative while in scatterable (free) states it remains positive at any separation.

An important result, giving us an idea of the size of the H atom, is the exact expectation (or the quantum mechanical average) value of the 'radius' <r> in a given state. For the ground state this is,

$$\langle r \rangle_H = 1.5 a_0 \tag{1.4}$$

Now consider the helium atom for which a detailed quantum mechanical treatment may be found in standard texts. Although the Schrödinger equation in this case does not yield an exact analytical solution, very accurate eigen functions and eigen values have been determined.

Some important points to note:

- Similar to the hydrogen atom the first two excited states of helium - denoted by 2^3S and 2^1S – have large excitation energies, while subsequent states are closer and closer together. The singly excited states of He, up to the first ionization threshold at $I_{He}= 24.6$ eV, are bunched within in an energy span of about 5 eV.
- The excited states 2^3S and 2^1S are metastable with large radiative lifetimes.
- The first ionization threshold of helium at 24.6 eV is the largest of any atom in the periodic table.

As per the scope of this book, suffice it to state that theoretical methods or structure calculations for atoms with atomic number $Z \geq 3$ began with the development of Hartree–Fock and Thomas–Fermi methods in the 1930s and currently may be calculated using commercial software package. Notable indeed was the work of Cox and Bonham (1967), who developed methods for determining the atomic electron charge density and static potential for atoms up to $Z = 54$ in parametric forms. Roothan–Hartree–Fock wave functions were presented analytically by Clementi and Roetti (1974) and were extended for heavier atoms with $Z = 55$–92 by (McLean and McLean 1981). Later improvements were made by Bunge and associates in 1993, and Koga and others in 1995.

1.3 ATOMIC PROPERTIES – ATOMIC RADII

Some of the well-known properties of atoms that we shall be exploring in this book are: the first ionization energy I; the average radius $<r>$; and the static electric dipole polarizability α_d measured as a volume quantity. In the case of the ground-state hydrogen atom, the values for I and $<r>$ are as already stated. For H-atom, α_d was obtained through a perturbation calculation (Schiff 1968) and is $\alpha_d = 9/2.a_0^3$ or 0.66 Å^3. One of the most accurately known properties of atoms is ionization threshold I, which shows systematic periodic variations as a function of Z, over the entire periodic table (Bransden and Joachain 2003). For hydrogen (or 1-electron) atom, one can combine equation (1.2) with the axiom that $<r> \sim n^2$, to derive the dependence of I on the average radius as $I \sim 1/<r>$. Assuming the polarizability $\alpha_d \sim <r>^3$, we arrive at a simple but useful conclusion,

$$\alpha_d \sim 1/I^3$$

Such a variation of the atomic polarizability with an inverse power of I has been discussed by several authors (Fricke 1986; Politzer et al. 1991; Dmitrieva and Plindov1983; Bhowmik et al. 2011). In order to test this relationship we have considered the inert gases and the halogen atoms in Figure 1.5. We have used atomic polarizabilities α_d available from the NIST Computational Chemistry Comparison and Benchmark Data Base (or the CCCBDB in short) and from Schwerdtfeger (2015). The plots of α_d versus $1/I^3$ are fairly linear with a very good correlation coefficient R^2. A good linear correlation is also observed for halogen atoms (Figure 1.5).

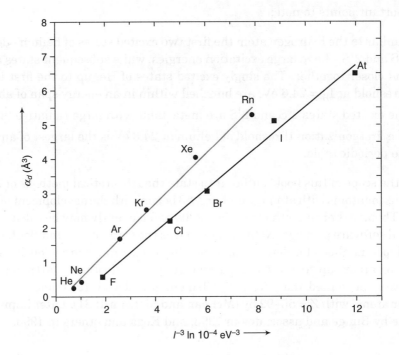

Figure 1.5   Linear correlation of α_d against $1/I^3$ for the inert-gas atoms and halogens with correlation coefficient $R^2 = 0.99446$ and 0.98975 respectively

A major focus is the *atomic/molecular scattering cross section* which is a dynamic, i.e., an energy dependent, area quantity. Therefore, it is necessary to introduce an elementary notion of *area* of an atom. The spatial extent of an atom can be gauged by defining its radius.

Quantum mechanics leads to the postulation of two types of radii. The orbital (or peak) radius 'r_p' is the radial distance from the nucleus to the peak position of the outermost orbital of electrons in the atom, while the average radius $<r>$ is the corresponding quantum mechanical expection value (Clementi and Roetti 1974; Bunge et al. 1993; Joshipura 2013). The Van der Waals radius R_W of two interacting identical atoms is defined to be the distance at which their inter-atomic attractive force turns repulsive. The fourth definition of radius comes from the response of the atom to an external electric field. Recall that the atomic electric dipole polarizability α_d is measured in volume terms. The polarizability radius is therefore defined as $R_{pol} = (\alpha_d)^{1/3}$ (Ganguly 2008). All the four types of radii of atoms are known across the entire periodic table, and the interesting behaviour of the atomic radii is shown in Figure 1.6, adopted from (Joshipura 2013; and references therein). The radii R_W and R_{pol} are shown here for inert gas atoms and alkali atoms only. For the helium atom the values (in Å) are: $r_p = 0.36$, $<r> = 0.49$, $R_{pol} = 0.59$ and $R_W = 1.4$.

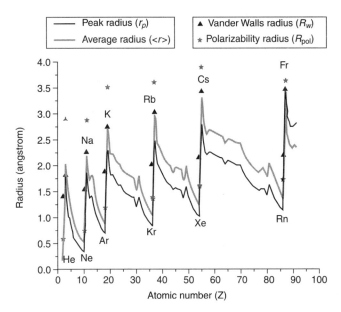

Figure 1.6   The four different radii r_p, $<r>$, R_W and R_{pol} of atoms across the periodic table (Joshipura 2013)

The graphical plot in Figure 1.6 shows hills and valleys in the atomic radii as functions of Z. Alkali atoms have the largest value in each peak with the caesium atom being the largest among them. This figure pictorially depicts the atomic radii in the ground atomic states. In the excited metastable states, the atomic sizes and polarizabilities are found to be relatively larger, and the ionization thresholds are appropriately smaller.

1.4   MOLECULAR STRUCTURE AND PROPERTIES

An amazingly wide variety of bound systems is created when atoms combine to form molecules. A molecule possesses a definite symmetry of its own that arises from its structure and bonding. These properties determine the designation of the electronic state of a molecule by a specific symbol, as illustrated in Box 1.2.

Molecules are characterized by important properties such as bond length(s), dissociation energies, ionization energies, dipole and/or quadrupole moment(s) and other parameters associated with rotational-vibrational motions. Moreover, in view of the anisotropic nature of these systems, it is required to define the average dipole polarizability (denoted by α_0). For molecules, a distinction is made between vertical and the adiabatic ionization energies. The (measured) vertical ionization energy corresponds to the resulting ion having the same geometry as the parent (neutral) species, while the adiabatic ionization energy is defined by considering a relaxed geometry of the ion. The ionization energies and other structure properties quoted in this book are experimental values, mostly from the well-known database CCCBDB. Dipole moments of polar molecules exhibit a wide range of values. One must mention here a unique hydrocarbon molecule methane, which is tetrahedral and close to spherical.

Table 1.1 Structural properties of a few common and small molecules (CCCBDB)

Molecule	Bond length (Å)	First ionization threshold (eV)	Dipole moment D (au)	Polarizability α_0 (Å³)
H_2	0.74	15.43	0	0.79
N_2	1.10	15.58	0	1.71
O_2	1.21	12.07	0	1.56
CO	1.13	14.01	0.044	1.95
NO	1.15	9.26	0.062	1.70
OH	0.97	13.02	0.60	0.98
CO_2	R_{C-O} = 1.16	13.77	0 (Linear geometry)	2.51
H_2O	R_{O-H} = 0.96	12.62	0.72	1.50
NH_3	R_{N-H} = 1.01	10.07	0.54	2.10
CH_4	R_{C-H} = 1.087	12.61	0	2.45

Polar as well as non-polar molecules play very significant roles in a variety of natural and man-made environments. The basic properties of atoms and molecules, highlighted in this discussion, are important because they provide input parameters for modeling the interactions of electrons with these systems. Table 1.1 highlights the structural properties of a few common and small molecules (from CCCBDB).

In atomic and molecular physics we often speak of a 'radical' which indicates a reactive atom or a molecule. Radicals are endowed with relatively low first ionization thresholds and high polarizability. For example, the CH, CH_2, and CH_3 radicals are more ionizable and polarizable than their parent stable molecule CH_4. Scattering of electrons by radicals presents challenges for both experiment and theory, as will be discussed later. Also, atoms and molecules which are highly ionizable and polarizable are more reactive, and offer relatively larger cross sections of scattering. Molecules in electronic metastable states form another class of interesting targets for electron scattering and may also have large electron scattering cross sections.

BOX 1.2: SYMBOLS FOR MOLECULAR STATES

Standard notations or the term symbols for electronic states of molecules are as follows: For a diatomic molecule, if the inter-nuclear axis is chosen as the Z-axis, the quantization of total electronic orbital angular momentum component L_z leads to projection quantum number $\Lambda = |M_L| = 0, 1, 2, 3 \dots$. Thus we have the following notations for the electronic state of a diatomic molecule.

$$\Lambda = 0, \quad 1, \quad 2, \quad 3$$

Symbol, $\qquad \updownarrow \quad \updownarrow \quad \updownarrow \quad \updownarrow$

$$\Sigma \quad \Pi \quad \Delta \quad \Phi \ \dots.$$

To describe an individual electron in the molecule, we use the corresponding notation for projection quantum number $\lambda = |m_l| = 0, 1, 2, 3 \dots$, and hence for an electron in a molecule,

$$\Lambda = 0, \quad 1, \quad 2, \quad 3$$

Symbol, $\qquad \updownarrow \quad \updownarrow \quad \updownarrow \quad \updownarrow$

$$\sigma \quad \pi \quad \delta \quad \phi \ \dots.$$

As an illustration, the electronic ground state of H_2 molecule is denoted by a complete notation $X^1\Sigma_g^+$. In this symbol, X stands for the ground electronic state, and the left superscript '1' indicates the spin singlet state. The '+' sign shows even symmetry of the electronic wave function upon reflection in the plane containing the nuclei and the subscript 'g' (gerade) indicates even symmetry with respect to inversion at the midpoint or center in a homonuclear diatomic molecule. In the case of odd symmetry in the inversion at midpoint, the symbol 'u' (ungerade) is used. One has to also specify the vibrational-rotational quantum numbers in a given electronic state.

Moreover, notations A, B, etc., are employed to specify the symmetry of the molecule in a given electronic state.

There have been several attempts to search for correlations or general trends in the properties among various classes of molecules and these are covered in later chapters. Efforts have also been made to express the molecular polarizability α_0 in terms of more accessible properties like number of electrons, bond length(s), ionization threshold, etc., and the dependence of α_0 on some inverse power of 'I' has been pointed out by Blair and Thakkar (2014). It seems, however, that unlike the case of selected atoms that have been examined in Figure1.5, there are no obvious trends for molecules.

～～～～～ ✪ ✪ ✪ ～～～～～

2 QUANTUM SCATTERING THEORIES

This chapter outlines the quantum scattering theories, employed later in this book, used for explaining the scattering of high energy (non-relativistic) electrons by atomic or molecular targets. The emphasis will be on a theoretical methodology based on the representation of the projectile-target system by a complex (optical) interaction potential. Electron scattering is measured (quantified) in terms of the 'cross sections' that result from the particular scattering event. As such, it is first necessary to define the different types of cross sections referred to, in this book, before further elaboration about the scattering processes.

2.1 DEFINITION OF ELECTRON SCATTERING CROSS SECTIONS

The probability of a scattering process is expressed in terms of a 'cross section', which is measured typically using the unit cm^2 or more conveniently in $\mathring{A}^2$.

The total cross section

The total cross section, $Q_T (E_i)$, describes the probability of the incident electron interacting with the target in a collision at a specific energy E_i and is defined by the total number of particles scattered in all directions per second per unit incident flux per scatterer. $Q_T (E_i)$ may be defined as the sum of individual cross sections for discrete scattering processes. Thus,

$$Q_T (E_i) = Q_{el} (E_i) + Q_{inel} (E_i) \tag{2.1}$$

where $Q_{el} (E_i)$ is the total elastic cross section and $Q_{inel} (E_i)$ is the summation of all inelastic scattering cross sections at incident energy E_i. Furthermore,

$$Q_{inel}(E_i) = \sum Q_{exc}(i \rightarrow f) + \sum Q_{ion}(A^{+n}) \tag{2.2}$$

where ΣQ_{exc} is the summation of cross sections for electronic excitation (from an initial state i to a final state f) and for molecules, also includes vibrational and rotational excitation. ΣQ_{ion} is the summation of cross sections for all ionization processes.

The elastic cross section

Elastic scattering conserves the total kinetic energy of the colliding particles. This means that quantum numbers that determine the energy are unchanged but other quantum numbers corresponding to degenerate states (e.g., helicity or spin flip) may change. In the case of many measurements not all states in the system are resolved. In this case 'effective elastic cross sections' are determined which may be referred to as 'rotationally unresolved, vibrationally unresolved, electronically unresolved', etc. Elastic cross sections are usually measured at specific energies and angles (the differential cross section). These data are used to determine the 'total elastic cross section' by integrating over the entire angular range (4π). The elastic differential cross section therefore provides a more stringent test of theory than a simple comparison with the total cross section. This is particularly true of collisions with polar molecules where empirical integral cross sections are often strongly dependent on the method used to represent low angle scattering. The total elastic cross section is given by

$$Q_{el}(E_i) = \int \frac{d\sigma(k;\theta,\varphi)}{d\Omega} d\Omega \tag{2.3}$$

where $\dfrac{d\sigma(k;\theta,\varphi)}{d\Omega}$ is the elastic differential cross section. The direction of incident electrons (see Figure 2.1) is chosen as the Z-axis, and the angle made by the direction of a scattered electron with Z-axis i.e., the polar angle is denoted by θ, while the corresponding azimuthal angle is denoted by φ. Thus (θ, φ) are referred to as the scattering angles. If $\mathbf{k}_i$ and $\mathbf{k}_f$ denote the incident and final (scattered) wave-vectors of the external electron then the elastic wave-vector magnitude is k = $|\mathbf{k}_i|$ = $|\mathbf{k}_f|$.

The momentum-transfer cross section

The elastic momentum-transfer cross section $Q_{MT.el}(E_i)$ is the $(1 - \cos\theta)$ weighted, angle-integrated differential cross section for elastic scattering.

$$Q_{MT.el}(E_i) = 2\pi \int \frac{d\sigma(k;\theta,\varphi)}{d\Omega} (1 - \cos\theta) \sin\theta \, d\theta \tag{2.4}$$

This can differ significantly from the total elastic cross section $Q_{el}(E_i)$, obtained by direct integration of the corresponding differential cross section over all angles. The momentum-transfer cross section may also be 'rotationally unresolved, vibrationally unresolved, electronically unresolved', etc. In electron transport modeling, input to the two-term electron Boltzmann equation requires that both $Q_{MT,el}$ and the total momentum-transfer cross section $Q_{MT,tot}$ (which includes the contribution of elastic, excitation and ionization processes) be such that $Q_{MT,tot} = \Sigma Q_{MTel} + \Sigma Q_{MTexc} + \Sigma Q_{MTion}$.

Inelastic cross sections

The total inelastic cross section $Q_{inel}(E_i)$, defined by equation (2.2), is the summation of all inelastic scattering cross sections at incident energy E_i. $Q_{inel}(E_i)$, is also defined as the total cross section $Q_T(E_i)$, minus total elastic cross section $Q_{el}(E_i)$ and may be evaluated theoretically. It may be described as the sum over cross sections where at least one quantum number that determines the energy is changed. $Q_{inel}(E_i)$ is often used in swarm/plasma experiments but is usually difficult to display using experimentally derived data (due to the error bars in total and integral elastic being larger than the summed inelastic cross section) in the evaluated data.

Some theoretical groups (e.g., Joshipura et al. 2017) report total inelastic cross sections for molecules at intermediate to high energies (typically from 15 eV onwards) as the sum of total ionization cross section ΣQ_{ion} and total electronic excitation cross section ΣQ_{exc}. This is justified since the contribution of rotational and vibrational excitation cross sections to the total inelastic cross section is negligible in this energy range.

The ionization cross section

The total ionization cross section, introduced in equation (2.2), describes the sum of all partial ionization cross sections $Q_{ion}(A^{+n})$ for atom A, where $n = 1$ stands for single ionization, $n = 2$ for double ionization, etc., and it includes inner shell ionization that can take place at high energies (>1 keV). In this text we will denote this sum simply by Q_{ion}.

2.2 DESCRIPTION OF EXPERIMENTAL MEASUREMENTS

Experiments on electron scattering form the basis on which quantum mechanical scattering theories are developed and hence it is necessary to have a brief review of a scattering experiment, leading to definitions of various cross sections.

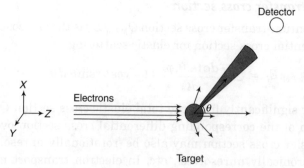

Figure 2.1 Schematic diagram of a scattering experiment; scattering angle θ is as shown here; the corresponding angle ϕ, not shown, is the azimuthal angle formed in the XY plane

In a typical laboratory experiment (Figure 2.1) a parallel mono-energetic beam of electrons is directed to impinge on the target (atomic or molecular beam) in a relatively small region called the scattering region. The electrons scatter from the target in different directions and are detected by a detector placed some distance (about a few centimetres) away. A steady beam of incident electrons continuously hitting the targets or scatterers, results in a steady stream of outgoing or scattered electrons, as if it were a steady picture. This phenomenon is appropriately described by the time-independent three-dimensional Schrödinger equation. The number of particles scattered in a particular direction per unit solid angle, per second, per unit incident flux, per scatterer is called the differential scattering cross section or (DCS) $\dfrac{d\sigma(k;\theta,\varphi)}{d\Omega}$. An elementary theory has been developed by solving the Schrödinger equation for scattering by a potential $V = V(\mathbf{r})$, with $r = |\mathbf{r}|$ as the radial coordinate of the projectile electron. An asymptotic boundary condition is imposed on the total scattering solution that defines the scattering amplitude $f(k; \theta, \varphi)$ (Joachain 1983, Bransden and Joachain 2003). The scattering amplitude representing the angular dependence of the outgoing spherical waves, is such that

$$\frac{d\sigma(k;\theta,\varphi)}{d\Omega} = |f(k;\theta,\varphi)|^2 \qquad (2.5)$$

Although this book is primarily focussed on theoretical analyzes of electron scattering from various atoms and molecules, the theoretical work should be compared with experimental data. As discussed earlier (in Chapter 1), the vast amount of electron scattering data needed for a diverse range of applications (covered later in Chapter 6) means it is not feasible for all such data to be generated experimentally. Indeed for many targets, such as radicals and reactive chemical species, experimental measurements are impractical. Experimental data is therefore used (both) directly, for providing cross sections for models and simulations, and as benchmarks for theoretical methods to establish the reliability of the particular method being used, while also ascribing an uncertainty to the theoretical data. In using experimental data it is necessary to understand the methodologies used and the errors associated with such data, including any systematic errors associated with these methodologies. Therefore, before we begin any detailed discussion of electron scattering processes and data relating to each target, it is necessary to review the experimental methods used to derive electron scattering cross sections.

Total scattering cross sections

The most basic electron scattering cross section is the total scattering cross section, a measure of the probability of an incident electron interacting with the target. The first experiments to derive total scattering cross sections were developed within a few years of the Franck–Hertz experiment (Franck and Hertz 1914) by Ramsauer in 1921. This has already been covered in Section 1.1.

The total cross section is derived using the Beer–Lambert law which states that if an incident electron beam, energy E_i, with a current I_{in} passes through a gas target with number density N, and path length D resulting in a transmitted current I_{out}, the total scattering cross section $Q_T(E_i)$ can be derived from the equation

$$I_{out} = I_{in} \exp\left(-N \cdot Q_T(E_i) \cdot D\right) \qquad (2.6)$$

In equation (2.6) N is determined by measuring the absolute pressure of the gas in the target cell.

There are, now, many variants of the Ramsauer apparatus with thermionic electron sources mostly replacing photo-cathodes, although Field and co-workers (Field et al. 1991) have developed a synchrotron based photoelectron source that conforms to Ramsauer's original experiment and is capable of exploring total cross sections at less than 10 meV. Supported by advances in pressure measurement, e.g., through use of calibrated baratrons, these experiments have measured $Q_T(E_i)$ for dozens of atomic and molecular targets with accuracies of 3–5%. These are some of the most accurate measurements of scattering cross sections. Nevertheless, such experiments may have some systematic errors; in particular, despite electron optics being used to collimate the incident electron beam, it remains difficult to distinguish between unscattered electrons and those 'scattered' at small angles (<5° to 10°). This may lead to larger errors (>10%) particularly for those molecular targets with dipole moments which favour scattering in the forward direction (small angle scattering). Such effects must be taken into account when comparing with theoretical data. These theoretical data may also be inaccurate in defining the dipole moment and scattering from such targets, particularly at low energies.

Electron impact ionization cross sections

The second most commonly (and accurately) measured electron scattering cross section is the electron impact ionization cross section, described by the process

$$e + A \rightarrow A^+ + 2e$$

It is relatively simple to detect ions – both positive ones (cations) and negative ones (anions) – since electronic counting of charged particles can be done with high (near 100%) efficiency. Accordingly, total electron impact ionization cross sections ΣQ_{ion} can be measured with small experimental/statistical errors and, with good pressure measurements, cross sections can be recorded with less than 5% uncertainty. The major systematic error is the collection efficiency of the ions from the interaction region (in case of crossed beams) or in the scattering cell. So, overall errors in the measured ΣQ_{ion} may be similar to those of $Q_T(E_i)$. In recent years, crossed beam configurations have often replaced older scattering cell geometries. With the need for careful calibration of ion collection efficiency, larger pulsed 'extraction' voltages are often more efficient but may suffer from poorer energy resolution. Figure 2.2 shows one apparatus used by Krishnakumar and co-workers (Krishnakumar and Srivastava 1988) to measure total electron impact ionization cross sections ΣQ_{ion}. Many of these have been compared with theoretical evaluations in later chapters of this book.

However, in the case of electron impact with molecules, ionization is not confined to direct ionization of the parent species; dissociative ionization can play a major part. Thus,

$$e + ABC \rightarrow A^+ + BC + 2e$$

$$\rightarrow A + BC^+ + 2e$$

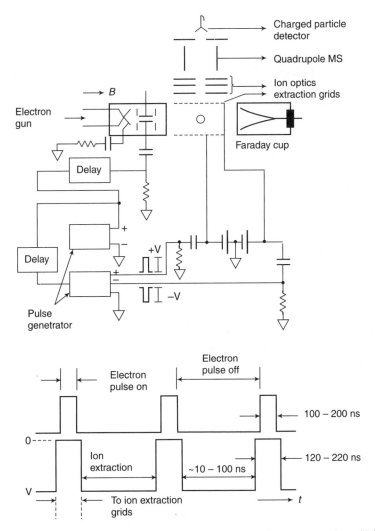

Figure 2.2 Apparatus used to measure electron impact ionization cross sections (Krishnakumar and Srivastava 1988)

The summation of the different *partial cross sections* for production of ionic fragments A^+, B^+, C^+, AB^+, BC^+, AC^+ with ABC^+ is the total ionization cross section ΣQ_{ion}. Measurements of partial cross sections require 'filtering' of the different fragment ions, usually by a mass spectrometric technique, introducing new experimental uncertainties.

The kinetic energy gained by the fragments in the dissociation process may lead to significant systematic errors particularly in crossed beam experiments where fast moving ionic fragments may overcome the ion collection potential and escape the interaction region leading to a lower cross section being recorded. This effect was often prevalent in early experiments, in fact until the 1980s, when it was duly recognized. Today the testing of the apparatus is usually conducted with greater care and corrected for errors. Such energy discrimination effects are also apparent in electron induced Dissociative Electron Attachment (DEA) experiments (covered below), since many product anions have residual kinetic energy. These systematic effects can lead to substantial apparent differences between theory and experiment (also discussed below).

Elastic scattering and differential scattering cross sections

The most commonly calculated electron scattering cross section is that of 'elastic scattering' $Q_{el}(E_i)$. In classical terms, elastic scattering means no energy is exchanged between the projectile and the target, though there may be exchange of momentum due to the change in trajectory of the incident particle. The angular distribution of the scattered particles is known as the *differential elastic scattering cross section* $\dfrac{d\sigma(k;\theta,\varphi)}{d\Omega}$, where (θ, φ) are the scattering angles (as depicted in Figure 2.1) and k is the elastic wave-vector magnitude. The Differential Scattering Cross Section (or DCS) is a more sensitive test of the scattering process, particularly at low energies (less than the ionization energy), since it is strongly influenced by the charge distribution of the target. In the case of a low energy incident electron, the target atom or molecule may be 'polarized' by the colliding electron and/or the incident electron may 'exchange' with an electron in the target. Accordingly, DCSs are a better test of the scattering code than the total cross section. Integration of the DCS over all angles (4π) allows the 'integral cross section' ICS to be determined which, below the first excitation threshold, should be same as $Q_T(E_i)$.

The need to test developing scattering theories and determine the mechanisms of electron scattering from atomic and molecular targets has led to the development of instruments capable of measuring $Q_{el}(E_i)$ and $\dfrac{d\sigma(k;\theta,\varphi)}{d\Omega}$. These instruments are now commonly known as 'electron spectrometers' since they can also explore excitation of target by measuring the energy lost by the incident electron during the collision. A typical electron spectrometer is shown in Figure 2.3 (Allan 2007, 2010). Electrons are commonly derived from a thermionic source (heated filament) and collimated and focused by a series of electron lenses. In order to produce an energy resolved electron beam capable of distinguishing elastic from inelastic processes, the electron beam is passed through a 'monochromator' which selects only a narrow (<100 meV) part of the thermionic source. This 'monochromated electron beam' is then focused into an interaction region where incident electrons scatter from the target. Scattered electrons are then collected, collimated, and focused into an 'electron analyzer' where elastic and inelastic scattered electrons are separated and only those for a specific process (elastic or excitation to a specific target state) are allowed through to the final detector. The analyzer is often rotatable to facilitate recording the DCS.

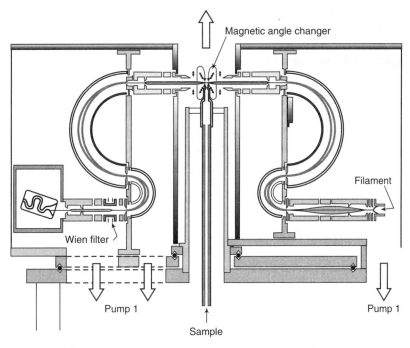

Figure 2.3 A typical electron scattering spectrometer (Allan 2007, 2010)

Such electron spectrometers have produced a large amount of scattering data and measured many DCSs for elastic scattering. Again, in order to produce accurate absolute scattering cross sections, it is necessary to carefully calibrate the spectrometer for systematic effects, and these include (Nickel et al. 1989):

- Monitoring of the incident electron current throughout often long periods of time, for collecting a complete DCS data set, usually by use of a Faraday cup in the interaction region.
- Transmission effects in the monochromator and analyzer.
- Careful calculation of the target density using the 'relative flow method'.

By such careful calibration, absolute elastic scattering cross sections at fixed energy and angle may be derived with accuracies of approximately 10% (though the quoted uncertainty is quite often higher).

Integration of the DCS to obtain integral cross sections (ICS) is common, but such data must be viewed with caution, because usually, DCS can only be measured between 20°/30° and 100°/120° due to the spatial limitations of the instrument. This can lead to large extrapolation errors particularly during forward scattering, where the DCS rises rapidly at low angles and errors of more than 30% in ICS are common. Recently the development of the 'magnetic angle changer' allows measuring DCS over the full angular range (0° to 180°) (King 2011) but this additional element is still absent in many currently operating electron spectrometers that are reporting data. Such ICS and $Q_T(E_i)$ may then be used to define the momentum transfer cross section $Q_{MT.el}(E_i)$ as in equation (2.4), albeit with similarly large uncertainties.

Inelastic scattering and resonance formation

Electron spectrometers operating in the 'inelastic scattering mode' are used to measure many inelastic scattering cross sections. The electron analyzer may be used to measure energy of electrons scattered from the target and record an 'Electron Energy Loss Spectrum' or EELS (Figure 2.4). Each energy loss peak corresponds to the transfer of incident kinetic energy to internal excited states of the target. In the case of atoms these define the atomic energy levels (Figure 2.4a), but for molecules the EELS is much

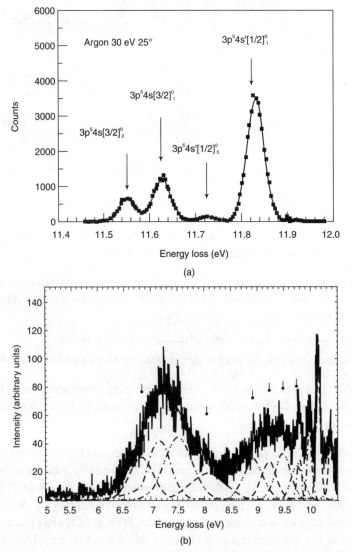

Figure 2.4 Electron Energy Loss spectra for (a) the Argon atom (Khakoo et al. 2004) and (b) the Water molecule (Thorn et al. 2007)

more complicated, with excitation not only of electronic states but also of rotational and vibrational states (Figure 2.4b). Usually, the energy resolution of the incident electron beam is insufficient to resolve any rotational states, so the measured 'elastic peak' includes a large number of unresolved rotational states in the molecular ground state. Hence, the measured cross section is known as 'quasi-elastic' and this should be remembered when comparisons are made with theory (theory being limited to defining a truly elastic cross section only when there is no rotational excitation). By tuning the analyzer to a particular energy loss, the DCS (and thence the ICS) of any excitation may be studied, leading to the derivation of excitation cross sections Q_{exc}.

ΣQ_{exc} is the summation of cross sections for electronic excitation and, in the case of molecules, includes vibrational and rotational excitation, and so may be used to determine the total inelastic scattering cross section Q_{inel} (E_i) defined in equation (2.2). Since for molecules many excitation processes may overlap (the ro-vibrational states of one electronic transition overlapping with those of another), it is often necessary to 'de-convolute' the measured EELS to obtain Q_{inel} (E_i). This is a complex process requiring many assumptions that can lead to major uncertainties. So much so, that 50% uncertainty is often the minimum that can be ascribed to such cross section estimates. Accordingly, such derived values may not be the best benchmark for theoretical calculations and it is now accepted that theory generated EELS should be compared with experimental data (Curik et al. 2015).

The development of 'monochromated' electron beams, i.e., beams with resolution of less than 100 meV, capable of resolving vibrational structure in energy loss spectra revealed unexpected fine structures in collision cross sections, such as a sharp dip in the elastic scattering cross section of electrons from helium at 19.317 eV (Schulz 1963) as illustrated by Figure 1.3 in Chapter 1. Such a phenomenon below the first excited state of helium at 19.819 eV appeared to have no explanation. But similar effects were seen in other atoms and molecules (Schulz 1973). Schulz proposed that these 'resonances' could be explained if the electron was temporarily 'captured' by the target to form a 'temporary negative ion' (TNI) with a life of only femto or picoseconds before decaying by auto-detachment, hence displaying delayed elastic scattering

$$e + AB \rightarrow AB^- \rightarrow AB + e$$

Subsequent experiments, particularly in nitrogen, revealed that this effect may also leave the target excited; indeed such phenomena may be the key factor in electron induced vibrational excitation. Resonance formation is a low energy phenomenon and so outside the scope of this book which focuses more on high energy electron scattering. However, because resonances are the major factor in particular in the electron scattering processes such as Dissociative Electron Attachment (DEA), they do deserve mention.

Dissociative electron attachment and dipolar dissociation cross sections

An alternative route of decay of the TNI results in the fragmentation of the molecule to form

$$AB^- \rightarrow A + B^-$$

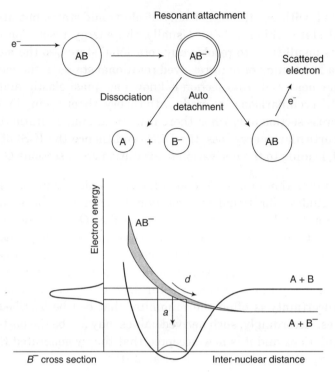

Figure 2.5 A schematic of the Dissociative Electron Attachment Process (DEA) with an illustration of the potential energy surfaces of neutral and temporary negative ionic states

This is a process called Dissociative Electron Attachment (DEA). First revealed in the 1960s, it is a process common to most molecules though the probability cross section is low ($< 10^{-18}$ cm^2) (Fabrikant et al. 2017). DEA cross sections are strongest for targets containing halogen atoms (since they are electronegative) but may be large in compounds containing 'pseudo-halogen' chemical groups such as –CN.

DEA may be readily explained by considering the potential energy surfaces of the anion and the neutral parent (Figure 2.5). The overlap of the anion curve and the neutral parent molecular potential energy curve determines both the cross section and the residual kinetic energy of the product anions. Since the anionic product is charged it can be readily detected by conventional mass spectrometry similarly to electron impact ionization. However, the kinetic energy released upon fragmentation may lead to significant systematic effects such as in measurement of H$^-$ from H$_2$. The H$^-$ fragments may have kinetic energies of several eV and thus escape the interaction region resulting in systematic lowering of the anion yield as a function of electron energy. such effects were responsible for discrepancies between experiment and theory for DEA from H$_2$ for more than three decades until Drexel et al. (2001) showed that these differences were a result of such kinetic energy discrimination effects amongst the measured anions. Although DEA is an important process in many natural and applied processes absolute DEA cross sections have only been measured for a few targets. Recently though, a new apparatus capable of measuring such cross sections with less than 10% uncertainties has been commissioned (May et al. 2010).

At higher energies direct fragmentation of the target may lead to formation of ion pairs

$$e + AB \rightarrow A^+ + B^- + e$$

The corresponding cross section is relatively small and is only a minor contributer to $Q_{\text{inel}}(E_i)$ or ΣQ_{ion}. Such a process has been studied for only a few targets (Szymańska et al. 2014) and there are few if any detailed theoretical calculations.

With this experimental background, it is back to the mainstream focus of this book, namely, theoretical calculations of electron scattering from atoms and molecules.

2.3 HIGH ENERGY ELECTRON SCATTERING

For the purpose of discussion in this book, the energy E_i of the incident electron will be considered as 'low' if it lies below the first ionization threshold 'I'; intermediate if it lies between the first threshold 'I' and about twenty times I; and if the energy $E_i > 20 . I$, it will be termed as high energy. The high energy domain assumes importance since a very large number of scattering channels open up in this region.

The Born and the related approximations

Soon after the advent of quantum mechanics, the Born approximation was developed to describe elastic scattering of fast particles by a potential $V(r)$. The approximation is, in principle, an infinite series of which the first term is the First Born Approximation (FBA). The FBA scattering amplitude is (Joachain 1983)

$$f_{B1}(k, \theta, \phi) = -\frac{1}{4\pi} \int \exp(i\Delta \cdot \mathbf{r}) \cdot U(\mathbf{r}) d\mathbf{r} \tag{2.7}$$

In this equation $\Delta = |\mathbf{k_i} - \mathbf{k_f}|$ is the elastic wave-vector transfer. The reduced potential is $U(\mathbf{r}) = \dfrac{2\,mV(\mathbf{r})}{\hbar^2}$, with m as the mass of the electron and $\hbar$ = (Planck's constant h)/2π. The energy of the incident electron is $E_i = \hbar^2 k^2/2m$. The Born series is a perturbation expansion in terms of $U(\mathbf{r})$. The FBA is a high energy, weak scattering approximation and further refinements, termed the second and the higher order Born amplitudes, have been made. A later development, called the Eikonal approximation, rests on an optical analogy and is an improvement over the FBA, in that it accounts for the distortion of the incident plane waves in the target region. A many-body formulation of the Eikonal approximation is called the Glauber approximation, which also exhibits term-wise expansion.

An effective high energy approximation called the '*Eikonal–Born series*' (EBS) theory was developed in the 1970s and 80s (Byron and Joachain 1977) to derive electron scattering cross sections more accurately, and was applied to hydrogen and other light atoms. The EBS direct elastic scattering amplitude f_{EBS}^d, without the electron-exchange effect, considered through $O(k^{-1})$ is expressed as follows

$$f_{EBS}^d = f_{B1} + \mathrm{Re}\, f_{B2} + f_{G3} + i\,\mathrm{Im}\, f_{B2} \tag{2.8}$$

In this four-term expansion, the subscripts B1 and B2 stand for the first and the second Born approximations, while f_{G3} indicates the third term of the Glauber approximation. Details of this theory along with the inclusion of electron exchange, through the high energy Ochkur amplitude g_{och}, are described in advanced textbooks on scattering theory (Joachain 1983, Bransden and Joachain 2003, Khare 2001). All of the four scattering amplitudes in equation (2.8) depend on the scattering angle θ through the elastic wave-vector transfer Δ. The imaginary second Born term included in the EBS expression (2.8) offers high energy evaluation of the total cross section through the optical theorem.

Before proceeding further it is useful to return briefly to the FBA in connection with the long-range electric dipole potential exhibited by a large class of molecular targets, namely, dipolar molecules. It is in this context that the FBA is significant. An external electron interacting with a polar molecule with a permanent dipole moment $\mathbf{D}$ ($D = |\mathbf{D}|$) experiences an attractive dipole potential

$$V_D \approx \frac{-e\, D \cos\beta}{r^2}, \text{ as } r \to \infty \tag{2.9}$$

where, e is the electron charge, $\mathbf{r}$ ($r = |\mathbf{r}|$) is the external electron coordinate and β is the angle made by $\mathbf{D}$ with $\mathbf{r}$. It is seen that V_D is an anisotropic long range point-dipole potential.

An extensive review on the collisions of slow electrons with polar molecules was published a few decades ago by Itikawa (1978). Polar molecules interact strongly with incident electrons and therefore the resulting low energy cross sections are often large. Several theoretical studies were carried out in the 1970s and 80s, on differential as well as total scattering of slow electrons by polar targets (Maru and Desai 1975, Joshipura and Desai 1980, and references therein). Recently Fabrikant (2016) has discussed, in a topical review, the scattering of slow electrons by polar molecules, including some modern perspectives.

The strength of the dipole potential is governed by the magnitude of $\mathbf{D}$. Classically speaking, the electric field of the external electron exerts a torque on the molecular dipole resulting in rotational motion. Quantum mechanically the electron excites the molecule from its initial rotational state J to the excited state $(J + 1)$. Rotational excitation energies are typically in the range of meV, and hence the rotational channels are open even at very low energies. Consider an electron with incident wave-vector $k_i = |\mathbf{k_i}|$ interacting with a weakly polar rigid rotator molecule having initial rotational quantum number J. Corresponding to the rotational excitation from initial state J to the final state $J' = (J + 1)$, the total cross section is given in the FBA by the following expression (Itikawa 1978)

$$Q_{rot}(E_i) = Q_{JJ'}(D, E_i) = \left(\frac{8\pi D^2}{3k_i^2}\right)\left(\frac{J'}{2J+1}\right)\ln\left|\frac{k_i+k_f}{k_i-k_f}\right| \tag{2.10}$$

The magnitude of the scattered vector is $k_f = |\mathbf{k_f}|$. The Born-dipole total cross sections provide an approximate but meaningful non-spherical contribution to the total electron scattering in appropriate cases. However, the simple form of the dipole potential, given in equation (2.9), only holds at large enough distances and diverges at the origin, and in the vicinity of the target molecule, the picture of an external electron interacting with the molecular point-dipole breaks down. Therefore, a more realistic model of the dipolar interaction can be represented as follows:

$$V_{DM} = -\frac{\vec{D}.\vec{r}}{r_d^3}; \quad 0 \le r \le r_d \tag{2.11}$$

$$= -\frac{\vec{D}.\hat{r}}{r^2}; \quad r \ge r_d$$

The first part of equation (2.11) shows the potential V_{DM}, zero at the origin, increasing linearly in magnitude with distance r up to a certain cut-off value r_d. Beyond $r = r_d$, this model reverts to the usual large r behaviour. This more realistic dipole cut-off model has been employed to calculate rotational excitation total cross sections (Shelat et al. 2011). Introducing (2.11) in (2.7), the first Born differential cross section is obtained in a closed form, but the total i.e., integral rotational excitation cross section Q_{rot} at a given energy is calculated through the numerical integration over the scattering angle θ (Shelat et al. 2011).

2.4 PARTIAL WAVE COMPLEX POTENTIAL FORMALISM

Another method for evaluating electron scattering cross sections uses Partial Wave Analysis (PWA). PWA requires the interaction potential $V = V(r)$ to be spherically symmetric. $V(r)$ is initially assumed to be real and the incident particles are treated as partial waves of different angular momenta $l = 0, 1, 2, 3, \ldots.$ etc., denoted by the usual symbols $s, p, d, f, \ldots$ etc. Thus one speaks of s-wave scattering, p-wave scattering and so on. In the PWA formalism, the scattering amplitude $f(k, \theta)$ is complex and reads as follows:

$$f(k, \theta) = \frac{1}{2ik}\sum_{l=0}^{\infty}(2l+1)[S_l(k)-1]P_l(\cos\theta) \tag{2.12}$$

In equation (2.12), δ_l is the phase-shift for l^{th} partial wave, $S_l(k) = \exp(2i\delta_l)$ is the S-matrix (scattering matrix) element and $P_l(\cos\theta)$ is the Legendre polynomial of order l. To evaluate the scattering amplitude given by equation (2.12) one needs the phase-shifts which are obtained by solving the radial Schrödinger equation, numerically in most cases of practical interest. The scattering amplitude in turn leads to the differential cross section and the integral (total) cross section denoted as $\sigma(k)$, which is purely elastic if the potential is real.

If the energy of the incident electrons exceeds the first excitation energy of the target (atom) then it is natural to expect inelastic scattering, wherein the incident particle loses energy, which goes into accessible excitations and ionizations of the target atom. At high enough energies many inelastic scattering channels are open together with elastic scattering. To account for this, the concept of a complex interaction potential has been evolved, analogous to the concept of complex refractive index of an absorptive medium, well-known in optics. The so-called complex 'optical' potential expressed as $V = V_R + iV_I$ (with V_R as the real part and V_I as the imaginary part) in the Schrödinger equation addresses a two-channel problem and accounts for elastic and cumulative inelastic scattering simultaneously. The important physical effects that occur when an electron interacts with a target can be incorporated in V_R. The imaginary part of the total potential, i.e., V_I, is usually referred to as the absorption potential V_{abs}.

The standard tool of partial wave analysis can still be used here if the total complex potential is spherical, and the term 'Spherical Complex Optical Potential' (SCOP) has been in common use for a long time. The complex potential yields complex phase-shifts represented by

$$\delta_l = \mathrm{Re}\delta_l + i\,\mathrm{Im}\,\delta_l \tag{2.13}$$

This means that the total (complete) cross section $Q_T(E_i)$ has to be the sum of total elastic cross section $Q_{el}(E_i)$ and cumulative inelastic cross section $Q_{inel}(E_i)$, as expressed in equation (2.1). The second term in equation (2.1) is also called 'absorption' cross section but absorption in the present context stands for removal or loss of particle flux into the inelastic channels and adheres to conservation of particle flux as well as energy. The standard equations (Joachain 1983) for the three total cross sections (TCS) referred to are:

$$Q_{el}(k) = \frac{\pi}{k^2}\sum_{l=0}^{\infty}(2l+1)\left|\eta_l\exp(2i\,\mathrm{Re}\,\delta_l)-1\right|^2 \tag{2.14}$$

$$Q_{inel}(k) = \frac{\pi}{k^2}\sum_{l=0}^{\infty}(2l+1)(1-\eta_l^2) \tag{2.15}$$

$$Q_T(k) = \frac{2\pi}{k^2}\sum_{l=0}^{\infty}(2l+1)[1-\eta_l\cos(2i\,\mathrm{Re}\,\delta_l)] \tag{2.16}$$

In these expressions, the quantity $\eta_l = \exp(-2\,\mathrm{Im}\,\delta_l)$ is known as the inelasticity or 'absorption factor' (Joachain 1983).

The phase-shifts and the TCS are evaluated by solving the Schrödinger equation. In an alternative approach developed by Calogaro (1967) the Schrödinger equation is transformed into a pair of coupled first order differential equations which lead to the scattering phase as a function of radial distance r, for a given l and k, and it stabilizes to the usual phase-shift δ_l at large r.

The cross sections Q_{el} and Q_T can be compared to the respective experimental measurements but in contrast, the inelastic TCS, $Q_{inel}(E_i)$ is not a single quantity. It is cumulative in the sense that it is the sum of the total (i.e., integral) cross sections representing all accessible excitation pathways and ionizations. Nevertheless the complex potential approach offers opportunities as well as challenges in electron collision studies, particularly when one looks at the theoretical quantity Q_{inel} in terms of its contents. In adopting the spherical complex 'optical' potential (SCOP) formalism we refer to the total projectile-target interaction as the complex scattering potential. For electron–atom scattering, the inelastic processes consist of transitions to discrete excited states and to continuum. Therefore, we bifurcate the total inelastic cross section as $Q_{inel} = \Sigma Q_{exc} + \Sigma Q_{ion}$, as stated in equation (2.2). Here, the first term represents the sum-total of the TCS of all energetically accessible excitations in the target atom from the initial state i to all accessible final states f. This term is also referred to as the excitation-sum. The second term in equation (2.2) is the sum of all total cross sections of energetically accessible ionization processes in the atom A, such that $n = 1$ stands for single ionization, $n = 2$ for double ionization, etc. At high enough energies inner shell ionization can also take place. Quite frequently in electron scattering studies, we refer to ionization processes, and for brevity the second sum in equation (2.2) will be denoted simply by Q_{ion}. The relative importance of the two terms in the said equation depends on the incident electron energy. Below the first ionization threshold, Q_{inel} consists entirely of excitation-sum ΣQ_{exc}. The thresholds of important low-lying excitations occur at relatively low energies compared to that of ionization. Ionization corresponds to transitions into continuum, which opens up infinitely many scattering channels. In general, at intermediate energies when the ionization cross section is increasing the excitation cross sections are decreasing and eventually Q_{ion} starts to dominate over the excitation-sum i.e., ΣQ_{exc} in equation (2.2).

Now, the crucial question: Can we deduce or extract the ionization TCS $Q_{ion}(E_i)$ from the known i.e., calculated $Q_{inel}(E_i)$? If yes, then we have a method of calculating the ionization cross section, together with an estimate of the quantity ΣQ_{exc}. A theoretical method for deducing Q_{ion} from Q_{inel} can be developed by introducing an energy dependent ratio

$$R(E_i) = \frac{Q_{ion}(E_i)}{Q_{inel}(E_i)} \tag{2.17}$$

The ratio function R seeks to determine Q_{ion} as a fraction of Q_{inel}. A sufficiently general semi-empirical method has been developed, over the last two decades, to determine the ionization TCS from the calculated inelastic TCS of a given target. The method called the *'Complex Scattering Potential – ionization contribution'* (CSP-ic) has been successfully applied to a large number of atomic as well as molecular targets (Joshipura et al. 2004, Antony et al. 2005, Vinodkumar et al. 2007).

Alternative methods for evaluating total ionization cross sections are the Binary-Encounter-Bethe or BEB approximation (NIST database and references therein) and the Deutsch–Maerk or DM formalism (Deutsch et al. 2000). The BEB model combines the Mott cross section with the high-incident energy behaviour of the Bethe cross section of electron scattering. The theory does not use any fitting parameters, and provides a simple analytic formula for the ionization cross section per atomic/molecular orbital. The total ionization cross section for a target is obtained by summing all these orbital cross sections. Four orbital constants, viz., the binding energy B, the orbital kinetic energy U, the electron occupation number N (see NIST database and references therein), and a dipole constant Q are needed for each orbital, the first three of which are readily available from the ground-state wave function of the target. The DM model expresses the total single electron-impact ionization cross section of an atom as a sum over contributions from atomic sub-shells (n,l), with each term constructed from the radius of maximum radial density $r_{n,l}$, corresponding number of electrons $\xi_{n,l}$ and appropriate weighing factors. Energy dependence in each term arises from reduced energy variable $u = E_i/E_{n,l}$, where $E_{n,l}$ is the sub-shell ionization energy. In another semi-empirical formulation electron–molecule ionization was investigated by Khare et al. (1999), combining useful theoretical features from Kim and Rudd (1994), and Saksena et al. (1997). This model also seeks to suitably combine the Bethe and the Mott cross sections as discussed at length by Khare (2001). The BEB, DM models and the method of Khare et al. (1999) focus only on Q_{ion}, while the CSP-ic method seeks to obtain Q_{ion} as a part of total inelastic cross section Q_{inel}. Atomic or molecular ionization induced by electron interactions has been a subject of many theoretical and experimental investigations, and many cross sections have been measured and derived. We therefore refer the reader to the NIST-USA database which lists electron impact cross sections for many atomic and molecular systems.

With this outline of different approximate theories complete, we will now discuss some results for the simplest atomic and molecular targets, namely H and H_2 respectively.

2.5 SCATTERING FROM ATOMIC AND MOLECULAR HYDROGEN

Atomic and molecular hydrogen constitute the most abundant species in the universe and therefore scattering of electrons from atomic or molecular hydrogen is important in many natural and technological processes. The purpose of this section is to see how the basic scattering theories outlined thus far can be applied to the simplest atomic and molecular targets H and H_2 respectively. Our discussion is confined mainly to intermediate and high energies.

Since pure beams of atomic hydrogen are more difficult to prepare there are still relatively few experimentally derived cross sections but in contrast there are many theoretical derivations. Total (complete) cross sections Q_T for the e–H, H_2 scattering in the energy range 1–300 eV were measured nearly twenty years ago by Zhou et al. (1997). For e–H scattering, electron impact BEB ionization cross sections and scaled

– Born excitation cross sections are available in the NIST database. Amongst a large variety of theoretical treatments on e–H scattering, a comprehensive review of the convergent close-coupling (CCC) theory on e–H excitation and ionization TCS appears in (Bray et al. 2012).

To briefly outline the FBA for e–H elastic scattering (see Joachain 1983), the electron charge density of the hydrogen atom is given exactly by $\rho_H(r) = \dfrac{1}{\pi} e^{-\lambda r}$ in au, where $\lambda = 2$ arises from the product of the ground-state wave functions. An atom, though electrically neutral, produces a static field and an electrostatic potential (energy). This, the so-called 'static' potential V_{st}, is given by the following expression for the e–H scattering system.

$$V_{\text{st}}(r) = -\left(1 + \frac{1}{r}\right) e^{-\lambda r}$$

The first Born elastic scattering amplitude is given by

$$f_{B1} = 2\frac{(\Delta^2 + 2\lambda^2)}{(\Delta^2 + \lambda^2)^2}$$

where, $\Delta = 2k\sin\left(\dfrac{\theta}{2}\right)$ is the elastic wave-vector transfer, k is the wave-vector magnitude of the incident electron, and θ is the scattering angle. This expression for f_{B1} yields the first Born differential cross sections DCS and the total (purely) elastic cross section for e–H scattering. An interesting result is that at high enough energies E_i, the Born elastic total cross section falls off as $1/k^2$ or $1/E_i$. For the Born total cross section of the dipole allowed 1s $\rightarrow$ 2p excitation, the high energy behaviour is lnE_i/E_i (see Joachain 1983).

The second term of the Born series expansion is known as the second Born e–H scattering amplitude f_{B2}. The complex quantity f_{B2} was evaluated with varying degrees of accuracy by various authors, e.g., Byron and Joachain (1977) and Byron et al. (1985). The total (complete) cross section, referred to in the earlier literature as the Bethe–Born cross section σ_{tot}^{BB} (Byron and Joachain 1977) is obtained by employing the imaginary term that occurs in the Eikonal–Born series (EBS) equation (2.8), in the optical theorem. Thus we have

$$\sigma_{\text{tot}}^{BB} = \frac{4\pi}{k} \text{Im} f_{B2}(\theta = 0) \tag{2.18}$$

Equation (2.18) represents the total (complete) cross section, in that it includes elastic plus all admissible non-elastic processes and it is $Q_T(E_i)$ in our present notation.

2.5.1 Electron collisions with atomic hydrogen

Results on Q_T for e–H scattering at three typical electron energies are shown in Table 2.1. Here, UEBS stands for Unitarized Eikonal–Born series, and SVP stands for

Schwinger Variational Principle (See Khare 2001 and references therein). The high energy Born-methods tend to overestimate the cross section at energies below 300 eV.

Table 2.1   Total (complete) cross sections Q_T (in Å^2) of e–H collisions at three energies, showing results of the EBS (Byron and Joachain 1977); the Unitarized EBS or UEBS (Byron et al. 1982); and the Schwinger Variational Principle – SVP

Energy E_i	EBS	UEBS	SVP	Other Data
100	2.10	1.97	1.85	1.75[a]
300	0.87	0.87	0.82	0.89[a]
1000	0.32	–	–	0.33[b]

Note: For EBS, UEBS and SVP data see Khare (2001); sources for other data are: a – measurements of Zhou et al. (1997), b – summed TCS vide equation (2.19) of the text.

Atomic hydrogen is a rather weak scatterer and a simple target in the sense that there are only a small number of important inelastic channels. Therefore, the total (complete) cross section for the H atom, which we denote by $Q_T(\text{H})$, incorporating the important scattering processes may be written explicitly as follows

$$Q_T(\text{H}) = Q_{el} + Q_{inel}$$

$$= Q_{el} + Q_{ion} + Q_{1s \to 2p} + Q_{exc} \tag{2.19}$$

In equation (2.19), Q_{ion} is the ionization TCS, while $Q_{1s \to 2p}$ denotes the TCS of the electron excited transition $1s \to 2p$. The last term Q_{exc} $(n > 2)$ is the sum-total of the excitation TCS for final states np with $n > 2$. The cross sections for optically forbidden transitions such as $1s \to 2s$ are fairly small. Figure 2.6 illustrates the relative contributions to the total cross section Q_T for atomic hydrogen, depicted at an incident electron energy of 60 eV in the form of a bar chart. 60 eV is the energy of the peak position of Q_{ion}, and the Q_T data are those measured by Zhou et al. (1997).

The TCS $Q_{1s \to 2p}$ and the $Q_{exc}(n > 2)$ determined in a scaled Born approximation are adopted from the NIST database. The Q_{el} for atomic hydrogen were calculated at various levels of theory (Khare 2001, and references therein). At this energy, the Q_{ion} and $Q_{1s \to 2p}$ (Figure 2.6) are approximately of the same magnitude, and contribute almost equally in the total cross section Q_T. This behaviour is observed at high energies too.

In general, it is possible that the individual TCS, i.e., different terms in the RHS of equation (2.19), obtained in experiment or in theory, may not add up exactly to the directly calculated or measured total (complete) cross section Q_T. If that is so, then we say that the consistency-check or *'sum-check'* is not exactly satisfied.

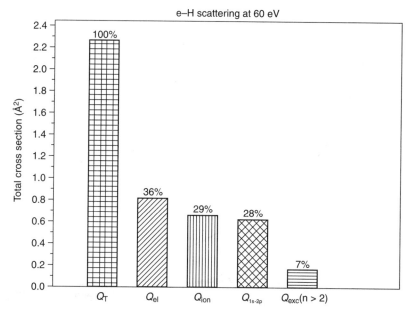

Figure 2.6 Bar chart showing the percentage contribution of various scattering processes in *e–H* collisions at the Q_{ion} peak of 60 eV, relative to experimental Q_T data (Zhou et al. 1997); Q_{el} from the sum-check in equation (2.19); the Q_{ion}, excitation TCS $Q_{1s \to 2p}$ and Q_{exc} (*n* > 2) from (NIST database)

In Figure 2.7, the quantities Q_T, Q_{el}, Q_{ion}, $Q_{1s \to 2p}$ and $Q_{exc}(n > 2)$ are plotted for atomic hydrogen as a function of incident electron energy over a wide range. The uppermost curve showing Q_T is based on the experimental data (Zhou et al. 1997) reported up to 300 eV, beyond which we have adopted the EBS results (Byron and Joachain 1977, Khare 2001). The experimental data and the EBS results on Q_T are seen to match at that energy.

At this stage, a brief review of the PWA calculations for atomic hydrogen is in order. An external electron approaching an atom experiences an electrostatic field or the static potential V_{st}, which is given by an exact expression in the case of *e–H* scattering. One must also incorporate another short range effect viz., that arising due to exchange of the incident electron with target electron. The effect can be included in the Schrödinger equation in the form of a local model potential V_{ex}. The exchange model adopted here (and in fact in all atomic molecular targets studied in this book) is the Hara free-electron–gas exchange model (Hara 1967).

An important long range interaction arises due to transient polarization of the target charge-cloud by the electric field of the external electron passing by and that is represented by the so-called correlation-polarization potential V_{cp}. The polarization effect is incorporated here through the correlation-polarization model potential V_{cp} (Perdew and Zunger 1981). At high energies, the polarization potential must be

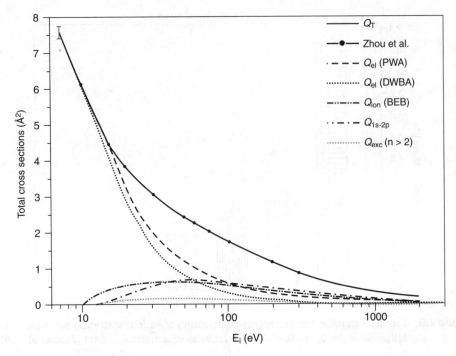

Figure 2.7 Major TCS (Å^2) of *e–H* scattering; uppermost curve: Q_T based on the measurements (Zhou et al. 1997) joined with the EBS values; next two curves: Q_{el} calculated in PWA and in DWBA (Benda and Karel 2014); lower three curves: Q_{ion}, $Q_{1s \to 2p}$, and $Q_{exc}(n > 2)$ from the NIST database

dynamic, i.e., energy dependent (Khare 2001). Thus, in order to obtain the elastic cross section Q_{el} of *e–H* collisions, a partial wave analysis of the total real potential, $V(r, E_i) = V_{st}(r) + V_{ex}(r, E_i) + V_{cp}(r)$ is required. Going back to results on atomic hydrogen, the calculated PWA elastic cross sections are depicted by the dashed curve in the middle of Figure 2.7. Comparison is made in this figure with the Q_{el} from the distorted wave Born approximation (DWBA) calculations of Benda and Karel (2014). The compared DWBA data are in accord with the previous theoretical results, as mentioned by the authors (Benda and Karel 2014).

Now consider the lower set of curves in Figure 2.7. The data for ionization and the excitation viz., Q_{ion}, $Q_{1s\to2p}$ and Q_{exc} ($n > 2$) plotted in this figure are those of the NIST database. To preserve clarity in this figure we have not shown other experimental data e.g., Q_{ion} of Shah et al. (1987) for hydrogen, but these are found to be in good agreement with the NIST ionization data.

The cross sections for *e–H* system as depicted in Figures 2.6 and 2.7 lead to several important conclusions. The first is that above the ionization maximum, the elastic scattering contribution declines relative to ionization as well as the 1s → 2p excitation. This means that at sufficiently high energies an incident electron is more likely to scatter from the H atom with, rather than without, energy loss. In terms of relative importance the 1s $\rightarrow$ 2p excitation competes with ionization even at high energies (Figure 2.7). An interesting observation is that in the case of atomic hydrogen, the ratio Q_{ion}/Q_{inel} is practically 0.5 at intermediate and high energies. H(2s) is a metastable state, but the optically forbidden transition 1s $\rightarrow$ 2s can be induced by electron scattering. This excitation results into the production of long-lived metastable H(2s) atoms in the ambient (gas) medium. Electron interactions with atomic hydrogen in its excited metastable state 2s, will be covered in Chapter 3. In contrast, electron induced excitation to the optically allowed 1s $\rightarrow$ 2p transition occurs with very large cross sections. As is well-known, H(2p) quickly de-excites by emitting the famous ultraviolet Lyman-α line.

Finally it is tempting to ask a general question; how large (or small) is the total cross section of atomic hydrogen, as compared to the physical 'size' or a representative cross sectional area of this atom (as discussed in Chapter 1)? The answer, of course depends, on the incident electron energy, since the cross sections are dynamic quantities. The measured values (Zhou et al. 1997) indicate that at 300 eV the Q_T of the Hydrogen atom is quite close to πa_0^2 (≈ 0.88 Å^2), being larger at lower energies and exhibiting a monotonous decrease as E_i increases (Figure 2.7). The ionization cross section for H atom reaches a maximum value of 0.66 Å^2 at about 60 eV (Figure 2.7) and then decreases slowly.

2.5.2   Electron collisions with molecular hydrogen

H$_2$ offers a simple prototype neutral molecular target, and can also be easily studied experimentally. In the following discussion we will not review the low energy collisions (below the ionization potential) which are discussed in a large number of publications, such as, Christophorou (2013), Tawara et al. (1990), Tennyson and Trevisan (2002), Anzai et al. (2012) and Yoon et al. (2008).

To begin with, consider the high energy domain wherein the de Broglie wavelength of the projectile electrons is smaller than inter-nuclear separation i.e., bond length $R = 1.4 \, a_0$ (0.74 Å) in H$_2$, so that the molecule can be viewed as a two-center system. At the dawn of quantum scattering theories in 1930s, an *'independent-atom model'* (IAM) was developed by Massey and others in UK (Mott and Massey 1965) and later on, after experimental data became available it was explored further in the 70s and 80s by theoretical groups in India (Jain et al. 1979, Jhanwar et al. 1980). High energy short wavelength electrons interact with two almost independent scattering centres (H atoms) in a target molecule H$_2$, bringing to mind Young's famous double-slit experiment on light. Indeed there is a revival of interest now in the molecular double-slit concept and electron interference in the impact ionization of the H$_2$ molecule as has been observed (Misra et al. 2004).

The H(1s) electron charge density in atomic units (au) is, $\rho_H(r) = (1/\pi)e^{-\lambda r}$, with $\lambda = 2Z$ and the atomic number $Z = 1$. For the H-atom bound in the H$_2$ molecule, the value of Z is variationally chosen to be $Z^* = 1.193$ (see Khare 2001) and this is to account for the covalent molecular bonding. In the IAM, the orientation-averaged elastic differential cross sections for high energy e–H_2 scattering denoted by $\bar{I}(\theta, k, Z^*)$, including the exchange effect, are expressed as follows (Khare 2001, Khare and Lata1985)

$$\bar{I}(\theta, k, Z^*) = 2\left|f_{\text{EBS}}^{\text{d}} - \frac{1}{2}g_{\text{och}}\right|^2\left[1 + \frac{\sin \Delta R}{\Delta R}\right] \tag{2.20}$$

In equation (2.20), the term $f_{\text{EBS}}^{\text{d}}(\theta, k, Z^*)$ is the direct EBS amplitude and $g_{\text{och}}(\theta, k, Z^*)$ is the Ochkur amplitude representing the electron exchange (Joachain 1983) for the H atom in H_2 molecule. The IAM formulation as in equation (2.20) takes into account the exchange effect in e–H_2 interactions (Khare and Lata 1985). Note the diffraction factor given by the square bracket expression in the above equation, which oscillates and tends to 1 at large wave-vector transfer Δ. The elastic e–H_2 DCS at energies above 100 eV are well reproduced by the IAM approximation (Khare 2001, and references therein).

Our interest in this book is in total cross sections. For molecular targets, a high energy estimation well known in literature is the Additivity Rule (AR) which seeks to approximate the molecular TCS as a sum of the TCS of the constituent atoms. Thus in a simple AR, valid at high enough energies, the TCS of a molecule AB is approximately the sum of the respective free-atom TCS, i.e.,

$$Q_T(AB) \approx Q_T(A) + Q_T(B) \tag{2.21}$$

This expression just ignores the *atom-in-molecule* aspects and, in general, tends to overestimate the cross section. Therefore a small correction is introduced in the simple AR for the molecular hydrogen case by calculating the atomic scattering amplitude and TCS, not for the free H atom, but for the one bound in H_2. Thus our total (complete) cross section of H_2 is given approximately through a modified AR and the optical theorem as

$$Q_T = 2\left[\frac{4\pi}{k}\text{Im} f_{B2}(\theta = 0; k; Z^*)\right] \tag{2.22}$$

A brief comparative study of high energy TCS Q_T for H_2 is shown in Table 2.2, which indicates that the modified AR, employed in equation (2.22), provides satisfactory agreement with the experimental data at high energies.

Table 2.2 Cross sections Q_T (in $Å^2$) of e–H_2 collisions at three energies, showing (a) modified AR as in equation (2.22), (b) the molecular EBS calculations (Khare 2001), (c) experimental data of Zhou et al. (1997), and (d) recommended data of Yoon et al. (2008)

Energy E_i eV	Theoretical results		Experimental/other data
	(a)	(b)	
100	3.31	3.22	2.56[c]
500	0.91	1.06	0.84[d]
1000	0.49	0.49	0.42[d]

At this stage, a few points are noteworthy in connection with molecular targets.

For molecules in general, the elastic TCS Q_{el} reported in the literature includes the rotational excitation TCS induced by the electrons. In other words, the Q_{el} are vibrationally elastic cross sections. For a non-polar target like H_2 rotational TCS are quite small (Christophorou 2013, Yoon et al. 2008, Anzai et al. 2012). In any case, for a molecule we must define the grand total cross section Q_{TOT} to include the non-spherical interactions, in the following manner

$$Q_{TOT}(E_i) = Q_T(E_i) + Q_{NS}(E_i) \tag{2.23}$$

where, Q_{NS} is the contribution due to total cross sections arising from rotational and vibrational excitations of the molecule. Equation (2.23) is most accurate when the non-spherical contribution Q_{NS} is rather small and in the present case of H_2 it is found to be reasonably accurate for incident energies above 10 eV. Therefore, it is useful to start with spherical electron–molecule interactions, calculate the total (complete) cross section Q_T and supplement this with an appropriate anisotropic contribution (determined separately) to describe the grand total cross section Q_{TOT}.

What does the TCS Q_{inel} actually represent? The molecular Q_{inel} represents the total cumulative inelastic contribution arising from electronic channels i.e., it includes all admissible electronic excitations as well as all ionization channels i.e., $Q_{inel} = \Sigma Q_{exc}(i \rightarrow f) + \Sigma Q_{ion}$, as stated earlier. Here, the first term includes the TCS for all accessible electronic transitions $i \rightarrow f$, where i is normally the ground electronic-vibrational state and f is the final electronic state for which the rotational-vibrational sublevels are governed by the Franck-Condon principle. The second term stands for the cumulative ionization TCS. It is important to note that the summation in the second term includes single, double and further higher levels of ionizations together with direct (parent) as well as dissociative ionizations and for the purpose of brevity, the summed total ionization cross section will be denoted just by Q_{ion} for a molecule. It is emphasized that if a realistic comparison of the Q_{ion} thus defined, is to be made with experimental molecular data, it is necessary to add all the contributions of single, double, direct and dissociative ionizations appropriately. For the hydrogen molecule an important observation is that the Q_{ion} is dominated by direct or parent ionization (Yoon et al. 2008) viz.

$$e + H_2 \rightarrow e + e + H_2^+ \tag{2.24}$$

It is possible to evaluate the family of five TCS $Q_T, Q_{el}, Q_{inel}, Q_{ion}$ and ΣQ_{exc} in the semi-empirical formalism called the 'complex scattering potential-ionization contribution' or CSP-ic method. This is an approximate method that starts essentially with complex potential representation of the electron-target system. Within the spherical approximation the four-term complex interaction potential is written explicitly as

$$V(r; E_i) = V_R + i\, V_I$$

$$= V_{st}(r) + V_{ex}(r, k) + V_{cp}(r, E_i) + i\, V_{abs}(r, E_i) \tag{2.25}$$

In the last expression, the imaginary term V_I is more appropriately denoted by V_{abs} and is the 'absorption' potential. Standard model potential forms are adopted to construct the full complex potential $V(r; E_i)$. Thus $V_{ex}(r, k)$ is from Hara (1967), and V_{cp} is from Zhang et al. (1992). The basic input required to generate the full complex potential, including V_{abs}, is the electron charge density of the H_2 molecule, in addition to some of its structure properties.

The radial charge density of the H_2 target is determined effectively as a function of distance r from the molecular center of mass, by a transformation involving the Bessel function expansion as discussed by Gradshteyn and Ryzhik (1965) and in other publications, e.g., Joshipura and Vinodkumar (1997). Starting with the charge density of the H atom (embedded in H_2) viz., $\rho(r, Z^*)$, and expanding it at the molecular center of mass, the single-center charge density $\rho_M(r; R, Z^*)$ of the H_2 molecule (having bond length R) is given approximately by,

$$\rho_M(r, R; Z^*) = 2\rho\left(r, Z^*; \frac{R}{2}\right) \tag{2.26}$$

In equation (2.26) r is the radial coordinate with respect to the mass-center of H_2. The required expression for the H-atom charge density is then (Joshipura and Vinodkumar 1997),

$$\rho(r, R) = \left(\frac{\lambda e^{-\lambda R}}{16\pi r R}\right)[(1 - r\lambda + R\lambda)e^{\lambda r} - (1 + r\lambda + R\lambda)e^{-\lambda r}] \quad \text{for } r \leq R$$

$$= \left(\frac{\lambda e^{-\lambda r}}{16\pi R r}\right)[(1 - R\lambda + r\lambda)e^{\lambda R} - (1 + R\lambda + r\lambda)e^{-\lambda R}] \quad \text{for } r > R \tag{2.27}$$

where, $\lambda = 2Z^*$. This expression can be used to determine the static potential $V_{st}(r, R/2)$ for the e–H_2 system. The charge density (2.27) is used as the input to construct all the other components of the e–H_2 complex potential $V(r; E_i)$. The single-centre expression for ρ_M also agrees closely with the accurate radial charge density of the molecule as derived from quantum chemistry codes, available as a part of the Quantemol-N scattering package (see Tennyson 2010, Tennyson et al. 2007).

Before we proceed further with scattering calculations, it is worthwhile to have a look at the first few electronic states of H_2 molecule and their respective excitation energies (Kim 2007, Yoon et al. 2008) as shown in Table 2.3. The ground electronic state of H_2 is $X^1\Sigma_g^+$ and hence the electron impact excitation to the first excited (repulsive or dissociative) state $b^3\Sigma_u^+$ must involve spin-exchange. Table 2.3 indicates that the important electronically excited states occur at energies below, but relatively closer to, the first ionization threshold (15.42 eV) of H_2, so that the energy gap between the excitation thresholds and the first ionization threshold is approximately 4 eV. This situation is similar to atomic hydrogen. Electron impact electronic excitation TCS for dipole allowed states are found in (da Costa et al. 2005), and for all the important electronic states of H_2 these are presented in (Yoon et al. 2008).

Table 2.3 The first few electronic states and their respective excitation thresholds of H_2 (Yoon et al. 2008); except for the data* for $b \, {}^3\Sigma_u^+$, which is from Anzai et al. (2012)

Electronic state	Vertical Excitation energy eV
$b \, {}^3\Sigma_u^+$	9.98*
$B \, {}^1\Sigma_u^+$	11.18
$c \, {}^3\Pi_u$	11.76
$a \, {}^3\Sigma_g^+$	11.79
$C \, {}^1\Pi_u$	12.29
$E, F \, {}^1\Sigma_g^+$	12.30
$e \, {}^3\Sigma_u^+$	13.23

Out of the vast amount of literature available on e–H_2 collision processes, we refer to the review paper of Yoon et al. (2008) and also to Brunger and Buckman (2002) in their extensive summary of electron molecule scattering. The measured Q_{TOT} in this case were determined by Hoffman et al. (1982), while experimental Q_{el} were measured by Khakoo and Trajmar (1986). Ionization cross sections of H_2, in its electronic ground state, were measured by Straub et al. (1996).

Against the backdrop of these data-sets, the total cross sections Q_T, Q_{el} and Q_{inel} of e–H_2 collisions were evaluated by Joshipura et al. (2010) in the important energy range from about 10 to 2000 eV using the SCOP formalism which starts with simultaneous elastic and inelastic electron scattering being represented by the spherical complex potential discussed above. A very useful and successful closed-form expression for a quasi-free Pauli-blocking absorption model to represent the potential $V_{abs} = V_{abs}(r, \Delta, E_i)$ was given by Staszewska et al. (1984). In this model, the inelastic interaction is viewed as the dispersion of the external electron in passing through the quasi-free electron-gas in the target atom or molecule. The absorption model potential of Staszewaska et al. reads as follows.

$$V_{abs} = -\rho(r) \sqrt{\frac{T_{loc}}{2}} \left(\frac{8\pi}{10 k_F^3 E_i} \right) \theta(p^2 - k_F^2 - 2\Delta)(A_1 + A_2 + A_3) \qquad (2.28)$$

where local kinetic energy of the external electron is obtained from

$$T_{loc} = E_i - V_r = E_i - (V_{st} + V_{ex} + V_p) \qquad (2.29)$$

Here, $\rho(r)$ is the electronic charge-density (ρ_M) of the target, and $p = 2E_i$ (in au) is the asymptotic incident momentum while $\theta(x)$ stands for the Heaviside unit-step function, whose value is zero for $x < 0$, and unity for $x > 0$. In the expression for local kinetic energy T_{loc}, the polarization potential V_p is almost insignificant. Furthermore $k_F = \sqrt[3]{3\pi^2 \rho(r)}$ is the Fermi wave vector magnitude. Detailed expressions for the three dynamic functions A_1, A_2 and A_3 are given in (Staszewska et al. 1984, Joshipura et al. 2010). In the model potential V_{abs} the energy parameter Δ is such that

$$V_{abs} = 0, \text{ for incident energy } E_i \leq \Delta$$

In that case no inelastic scattering can occur. Therefore in principle Δ is the first electronic excitation threshold of the target. However, the first excitation threshold is usually small and for such low values of Δ the model absorption potential is rather large and it results in a rather large removal of flux into inelastic channels, and therefore an appropriate choice of this parameter becomes necessary on physical grounds (Joshipura et al. 2010). A reasonable choice would be to fix Δ at the first ionization threshold. Let us briefly mention here that a variable form of the parameter Δ is found to be more suitable, as discussed by Joshipura et al. (2010). Further details of the potential models employed in CSP-ic are also described in Joshipura et al. (2004, 2009). In summary, the TCS Q_T, Q_{el} and Q_{inel} for H_2 target were calculated using the CSP-ic, (Joshipura et al. 2010) while, the grand TCS Q_{TOT} were obtained through equation (2.23), i.e., by 'including' the non-spherical (rotational excitation) TCS contributions, available from Yoon et al. (2008).

Our next task is to deduce the ionization cross section within Q_{inel}. The cumulative TCS Q_{inel} starts from an appropriate threshold, rises to a maximum value at an intermediate energy which we will denote by E_P, and falls off slowly at higher energies. Q_{ion} appears at its own threshold i.e., 'T', rises to its maximum at an energy somewhat higher than E_P, and falls off rather slowly with energy, such that

$$Q_{ion} \lesssim Q_{inel} \qquad (2.30)$$

For common atomic and molecular targets (but not for the H atom), a general trend is that towards high energies the Q_{ion} tends to be closer and closer to Q_{inel}. This is expected since the important excitation channels open up at thresholds lower than 'T' and the corresponding cross sections Q_{exc} begin to decrease with energy, when the Q_{ion} increases and dominates in Q_{inel}. The aforesaid qualitative description can be cast into mathematical conditions on the ratio function

$$R(E_i) = Q_{ion} / Q_{inel}$$

Thus (Joshipura et al. 2004, 2009),

$$R(E_i) = 0, \text{ at and below } E_i = I \tag{2.31a}$$

$$= R_P, \text{ at an intermediate energy } E_i = E_P \tag{2.31b}$$

$$\lesssim 1, \text{ at high energies } E_i \gg E_P \tag{2.31c}$$

A rigorous assessment of the intermediate and high energy behaviour indicated in equation (2.31b, c) does not seem to be possible. However, for the H_2 molecule in particular, a reasonably good estimate is possible in view of the amount of reliable data available. Thus one can construct a 'reference data-set' for Q_{inel} of e–H_2 scattering by combining the literature results on ionization (Straub et al. 1996) and electronic excitations (Yoon et al. 2008, da Costa et al. 2005). Specifically, Joshipura et al. (2010) combined the individual excitation TCS Q_{exc} of the first few electronic states $b^3\Sigma_u^+$, $B^1\Sigma_u^+$, $c^3\Pi_u$, $a^3\Sigma_g^+$, $C^1\Pi_u$, E, $F^1\Sigma_g^+$ and $e^3\Sigma_u^+$ adopted from the recommended results (Yoon et al. 2008) to generate the reference values of the quantity ΣQ_{exc}. To this electronic-sum one can add the recommended Q_{ion} (Yoon et al. 2008) to yield a reference data-set for Q_{inel} at various energies. The purpose (Joshipura et al. 2010) was to obtain for H_2, the above mentioned ratio at intermediate and high energy, vide equation (2.31b, c). For the hydrogen molecule the ratio $R(E_i)$ is found to be almost 0.5 at the Q_{inel} peak $E_P (= 50 \text{ eV})$, and this suggests equal contributions from ionizations and electronic excitations at that energy. The reference data-set also indicates that the ratio function $R(E_i)$ increases steadily to about 0.7 at 700 eV.

In summary the ionization fraction $R(E_i)$ is zero at the threshold I, achieves a value 0.5 at the intermediate energy E_P and continues to rise slowly to almost 0.7 at a high energy of 700 eV. With the three conditions thus obtained for H_2 it is possible to represent R as a continuous function of energy. We express incident energy as a target-specific scaled variable $U = E_i/I$, and write the ratio function parametrically as follows (Joshipura et al.)

$$R(E_i) = 1 - C_1 \left[\frac{C_2}{U+a} + \frac{\ln U}{U} \right] \tag{2.32}$$

The three-parameter equation (2.32), though not unique, is constructed such that the R approaches unity as the second term approaches zero at high energies. The reason for choosing a particular expression for R is the following. At high enough energies, discrete excitation cross sections dominated by dipole transitions tend to fall off as $\ln U/U$ and accordingly the decrease of the second term in equation (2.32) must also be proportional to $\ln U/U$. However, one more term viz., $C_2/(U+a)$ has been introduced here, so as to ensure a better energy dependence at low and intermediate energies. We can then employ the three conditions (2.31a-c) to evaluate the dimensionless parameters a, C_1 and C_2 introduced in equation (2.32). This procedure enables us to determine the Q_{ion} from the calculated Q_{inel} through the function R expressed in equation (2.32). Table 2.4 shows the values of these parameters for H_2.

Table 2.4   Properties and CSP-ic parameters of H_2 (Joshipura et al. 2010)

I in eV	E_P *in eV*	*a*	C_1	C_2
15.42	50	−0.2756	0.081	6.26

Equations (2.17), (2.30), (2.31a-c) and (2.32) define the CSP-ic method as employed by Joshipura et al. for the hydrogen molecule. It enables us to deduce not only Q_{ion} but ΣQ_{exc} as well from the Q_{inel} calculated in SCOP.

The CSP-ic ionization cross sections and the electronic-sum for molecular hydrogen are shown in the all-TCS plot Figure 2.8, which offers a detailed comparison of the calculated cross sections with a number of other theoretical and experimental data. The two top most curves in this figure are the grand total cross sections Q_{TOT} for which the calculated values (Joshipura et al. 2010) are in a good general agreement (above 20 eV) with the recommended data (of Yoon et al.) and measurements of Hoffman et al. (1982). At lower energies, the CSP-ic Q_{TOT} tends to fall below the other compatible data, perhaps, in part due to inadequate inclusion of the non-spherical contribution Q_{NS}. The recommended grand total cross sections Q_{TOT} in (of Yoon et al.) are the experimental data. Also included in the graphical plot (Figure 2.8) are the CSP-ic Q_{el}

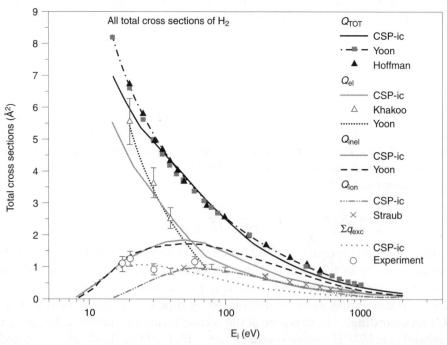

Figure 2.8   Electron scattering cross sections ($\mathring{A}^2$) with H_2 molecules; Q_{TOT}, *Solid line*: CSP-ic; (Joshipura et al. 2010); *solid dash dot line with squares*: recommended data (Yoon et al. 2008); *solid triangles*: Hoffman et al. (1982); Q_{el}, *solid line*: CSP-ic; *open triangles*: Khakoo and Trajmar (1986); *dots*: Yoon et al. (2008); Q_{inel}, *solid line*: CSP-ic; *dash reference data-set (see text)*; Q_{ion}, *'dash double dot'*: CSP-ic; *crosses*: measured data – Straub et al. (1996); ΣQ_{exc}, *dots*: CSP-ic and *open circles*: from the reference data-set (see text)

along with other results. CSP-ic Q_{inel} are seen to be in agreement with those generated from the 'reference data-set' mentioned above, up to about 100 eV. One also notices that the Q_{el} curve crosses the Q_{inel} curve at the peak (E_p) of Q_{inel}. This indicates equal amounts of elastic and cumulative inelastic scattering at the peak position E_p and that is expected on theoretical grounds (see Bransden and Joachain 2003).

The elastic TCS falls off faster and tends to be closer to the ionization TCS at high energies. One also finds (Figure 2.8) that above 100 eV, the CSP-ic values of Q_{inel} are lower than the reference data-set, with the latter decreasing rather slowly and approaching the Q_{TOT} values. That in turn, renders the elastic counter-part Q_{el} to be very small, with the result that the sum-check for grand total cross sections would not be satisfied. Therefore, it appears that the reference data-values Q_{inel} deduced as above are overestimating the cross section for energies exceeding 500 eV.

As regards the ionization results on H_2, we can see in Figure 2.8 that the CSP-ic Q_{ion} curve rises somewhat faster from the threshold, and matches with the ΣQ_{exc} at the inelastic peak $Ep = 50$ eV. One may recall here the case of atomic H for which the ionization and excitation contributions are also found to be nearly the same (Figure 2.7) at the Q_{ion} peak. Finally in Figure 2.9 CSP-ic derived ionization cross sections are compared with the recommended data of Yoon et al. and BEB derived values (NIST database).

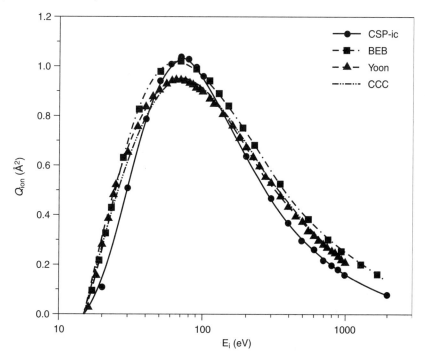

Figure 2.9 Total ionization cross sections in e–H_2 scattering; CSP-ic results compared with the recommended data (Yoon et al. 2008), BEB theory (NIST database), and CCC theory (Zammit et al. 2016)

It is interesting to note at this stage that an elaborate complete solution of the e–H_2 scattering problem using the accurate target states in the CCC formulation has been discussed recently by Zammit et al. (2016). The Q_{ion} results of (Zammit et al. 2016) together with the other results displayed in Figure 2.9 appear to be within experimental uncertainties of ~15%. An interesting feature of e–H_2 ionization noticed from the recommended data of Yoon et al. is that the Q_{ion} are greatly dominated by parent ionization i.e., production of H_2^+. At the peak of the ionization cross section around 70 eV, the parent ionization TCS is almost 93% of Q_{ion}.

One of our objectives throughout this monograph is to study the relative variations of different TCS of electron–atom/molecule scattering with respect to the incident energy. Figure 2.8 presents, in this respect, an overall picture of the important TCS of e–H_2 collisions in the energy range considered. From the CSP-ic results at 100 eV, we conclude that $Q_{TOT} = 2.1$ Å^2, while $Q_{el} = 0.85$ Å^2 and $Q_{ion} = 0.96$ Å^2. At higher energies, the grand total cross sections Q_{TOT} of this molecule are composed mainly of Q_{el} and Q_{ion}, and it is found that the ratio Q_{ion}/Q_{TOT} tends to be close to 0.5.

Before we close the discussion on H_2, it is of interest to mention briefly the approximate calculations reported in (Pandya and Joshipura 2010) on a particular excitation process important at low energies, and that is the production of neutral Hydrogen atoms in e–H_2 inelastic scattering. Consider the electron impact neutral dissociation of this molecule, i.e.,

$$e + H_2\ (X^1\Sigma_g^{\ +}) \rightarrow H_2^*(b^3\Sigma_u^{\ +}) \rightarrow H\ (1s) + H\ (1s) + e \tag{2.33}$$

A theoretical calculation of the total cross sections for the above process was carried out in (Pandya and Joshipura 2010, 2014) by using the R-matrix Quantemol-N scattering code (see Tennyson 2010, Tennyson et al. 2007). In the said work, Pandya and Joshipura included appropriately the cascading from the electron excited higher triplet states in H_2, over and above the basic excitation-dissociation process represented by equation (2.33). Moreover, the process (2.33) releases two H atoms per incident electron against which a single H atom is produced in the electron impact dissociative ionization of H_2 molecule, taking place at high enough energies. Thus to make a better estimate on the production of neutral H atoms in e–H_2 scattering, we must also consider dissociative ionization, i.e.,

$$e + H_2\ (X^1\Sigma_g^{\ +}) \rightarrow H\ (1s) + H^+ + e + e \tag{2.34}$$

Detailed results are not given here, but the maximum magnitude of Q_{Ndiss} is found (Pandya and Joshipura 2010) to be close to 1.1Å^2 at energy of about 15 eV. This value is close to the maximum neutral dissociation cross section of 0.9 Å^2 at 16.5 eV with an uncertainty of 20%, as quoted in the data paper of Yoon et al. (2008). An interesting aspect of the neutral dissociation of H_2 is that a very small fraction of H atoms produced is likely to be in metastable state 2s.

It has been emphasized frequently in literature that no single theory can produce electron impact cross sections consistently from low to high energies. However, a suitable combination of two or more theoretical formulations can be made and this was done, for example by Korot et al. (2012) in which the e–H_2 TCS Q_{TOT} were obtained over a wide energy range from 0.01 eV to 2 keV. In the work of Korot et al. (2012) a hybrid model was employed. Low energy Q_{TOT} were generated by using the R-matrix Quantemol-N scattering code up to 15 eV – close to the first ionization threshold of the molecule – beyond which, the spherical complex potential method was used. There was, more or less, a smooth matching of the theoretical results, at 15 eV, in transferring from one theory to the other.

2.6 CONCLUSIONS

In this chapter the basic scattering theories have been outlined and e–H, H_2 interactions have been discussed as simple applications of these theories. High energy Born-type approximations are shown to be of only limited use, but these approximations provide useful theoretical tools to examine electron scattering in specific situations such as external plasma, and this has been demonstrated in several models of technological plasmas (e.g., Modi et al. 2015). The CSP-ic method has also been developed and discussed in this chapter for the specific case of H_2; this method will be generalized to other atomic and molecular targets in the subsequent chapters.

It should be noted that in the above discussion, we have been able to cover only a few of the many phenomena that take place in the interactions of fast electrons with hydrogen molecules. In general, low energy phenomena are not considered in this book. Yoon et al. (2008) addressed, in their data paper on H_2, about a decade ago, some of the outstanding issues and had suggested future problems on e–H_2 molecule collisions. One of these was a suggestion to carry out investigation on the electron scattering from electronically excited metastable states of this famous and fundamentally significant molecule. A theoretical investigation on H_2 in its metastable state has been conducted subsequently using the CSP-ic and it will be discussed later in this book in Chapter 4.

3 ELECTRON ATOM SCATTERING AND IONIZATION

The basic theories outlined in Chapter 2 on the scattering of electrons from atomic targets other than H, will be applied in this chapter. Formulating a theory of electron scattering by atomic targets is in a way simple since the center of mass of the target-projectile system remains essentially fixed in the target while it is the outgoing electron that suffers momentum (and possibly energy) transfer. Applying the Born and the related approximations to many-electron targets becomes increasingly difficult, and several better alternatives have emerged in the last few decades and these will be discussed, as appropriate.

3.1 INTRODUCTION

At incident electron energies (E_i) below the first excitation threshold E_{ex} of the target, the collisions are purely elastic and the elastic cross section is the same as the total (complete) cross section Q_T. At energies above E_{ex} there is simultaneous elastic and inelastic scattering and once again,

$$Q_T(E_i) = Q_{el}(E_i) + Q_{inel}(E_i)$$

where $Q_{el}(E_i)$ is the total elastic cross section and $Q_{inel}(E_i)$ is the summation of all inelastic scattering cross sections at incident energy E_i. Furthermore,

$$Q_{inel}(E_i) = \sum Q_{exc}(i \to f) + \sum Q_{ion}(A^{+n})$$

A list of atomic targets included in this chapter, in a rather arbitrary sequence, is given in Box 3.1.

Earlier reviews of electron–atom collision studies may be found in Zecca et al. (1996) and Khare (2001). Remarkable progress has been made since then, especially in theory. A viable theoretical tool for the treatment of scattering employed at intermediate and high energies is the method of complex potential partial wave expansion. Electron–atom ionization forms a part of various scattering processes described in this formalism. The methodology rests on the representation of the projectile-target system by the total spherical complex potential $V_{tot} = V_{st} + V_{ex} + V_{cp} + i\, V_{abs}$, as outlined in the previous chapter. For an incident electron all the four terms of the complex potential are attractive. The complex potential approach has also come to be known as optical model for calculations. In a target atom, the electron cloud screens the nuclear charge and gives rise to the short range static potential $V_{st}(r)$, where r is the radial coordinate of the external electron. The possibility of exchange between the incident electron and one of the target electrons gives rise to the short-range exchange potential V_{ex}, e.g., Hara (1967). The incident electron also induces distortion or polarization effects in the target charge-cloud. At large distances from the atom, the polarization effect is represented by a long-range potential $-\alpha_d/2r^4$, with α_d as the static atomic dipole polarizability. This simple form is asymptotic and is inadequate at short distances, therefore a better representation, known as the correlation–polarization model potential $V_{cp}(r)$ has been developed in literature. This model seeks to combine the long range polarization term $-\alpha_d/2r^4$ smoothly with the short-range correlation term $V_{co}(r)$. An effective correlation–polarization model in this regard is that of Zhang et al. (1992), expressed as

$$V_{cp} = -\frac{\alpha_d}{2(r^2 + r_{co}^2)^2} \tag{3.1}$$

BOX 3.1

Atomic targets for which electron scattering cross sections are presented in this chapter–

Inert Gases: Helium (He), Neon (Ne), Argon (Ar), Krypton (Kr) and Xenon (Xe)

Prominent Atoms: Oxygen (O), Nitrogen (N) and Carbon (C)

Other Important Atoms: Beryllium (Be) and Boron (B)

Halogens: Flourine (F), Chlorine (Cl), Bromine (Br) and Iodine (I)

Less Studied Atoms: Silicon (Si), Phosphorus (P), Sulphur (S) and Germanium (Ge)

Important Metals: Aluminium (Al) and Copper (Cu)

Metastable Atomic Species: H* & He*; O* and N*

The correlation parameter r_{co} involved in equation (3.1) is determined through the following condition (Zhang et al. 1992),

$$V_{cp}(0) = -\frac{\alpha_d}{2r_{co}^4} = V_{co}(0) \tag{3.2}$$

The Zhang polarization potential (3.1) smoothly approaches the correct asymptotic form at large r.

The imaginary part of the spherical complex optical potential (SCOP) is the so-called 'absorption' potential $V_{abs} = V_{abs}(r, \Delta, E_i)$. The basic form of V_{abs} currently adopted by many modellers, for instance, Blanco and Garcia (1999), was formulated by Staszewska et al. (1984). Detailed expressions of V_{abs} have been given by equation (2.28) and equation (2.29) in Chapter 2. The absorption potential depends on a parameter labelled as Δ. This is a threshold energy parameter of the target with respect to incident electron, such that $V_{abs} = 0$, if $E_i \le \Delta$. Ideally Δ is the first excitation threshold E_{ex} but it may be conveniently replaced by the first ionization energy I.

The SCOP approach is found to give reliable data for many targets but the method should be treated with caution. The approach itself may not be reliable at low energies, e.g., below the ionization threshold and often shows discrepancies with experimental data below 50 eV. Also the inelastic channels are represented cumulatively, not individually, and hence the corresponding cross section that one derives by using this methodology is the summed total inelastic cross section Q_{inel}. This quantity, together with total elastic cross section Q_{el} leads us to the total (complete) atomic cross section Q_T.

Of great interest amongst the inelastic channels (at energies above the threshold 'I') is ionization but it is implicit in this formalism, in the sense that the corresponding summed total cross section (TCS) Q_{ion} is included in Q_{inel}, i.e., $Q_{inel} = \Sigma Q_{exc} + \Sigma Q_{ion}$. Here, the first term is the sum of total electronic excitation cross sections from initial state i to final states f. The second term is the cumulative total ionization cross section Q_{ion} representing single, double, and further levels of ionization of the target atom. Since the threshold for double ionization is higher than that for the single ionization, the TCS for single ionization $Q_{ion}(A^+)$ for atom A is much larger than that for double ionization viz., $Q_{ion}(A^{+2})$. In general we have $Q_{ion} \lesssim Q_{inel}$.

The '*Complex Scattering Potential – ionization contribution*' (CSP-ic) method is a semi-empirical approximation based on general quantum mechanical considerations and has been described in Chapter 2. CSP-ic has achieved a reasonable degree of success in calculating the total ionization cross section Q_{ion} across a wide variety of atoms and molecules, correctly reproducing the shapes and magnitudes of ionization cross sections. In most of the cases the derived cross sections are within the uncertainties (usually about 10 to 15%) generally observed in the experimental ionization data.

In the following sections, cross sections for electron scattering from selected atomic targets are presented and compared with available experimental and/or theoretical results, including, in a few cases, published recommended data. The sections are arranged in terms of certain groups of atoms.

3.2   INERT GAS ATOMS

Electron interactions with inert gases are relevant in studying the atmospheres of some planets, in producing population inversion in laser systems (e.g., the He–Ne laser) and in plasmas used in modern industry as a tool for surgery and dentistry. Electron scattering from inert gases, such as He, Ne, Ar, Kr and Xe, has traditionally been used to provide standards for developing both theoretical and experimental methods; with the observation of the Ramsauer–Townsend effect (also known as RT minimum) in the total cross section of electron scattering from inert-gas atoms being an excellent benchmark. The RT minimum makes these atoms appear 'transparent' to the incident electron. The RT minimum is now interpreted by considering a partial wave analysis, vide equation (2.12) of Chapter 2. At a low energy (typically below 1 eV), the s-wave scattering is dominant such that when the corresponding phase-shift $\delta_0 = \pi$ the resulting cross section dips to a minimum.

3.2.1 He and Ne

We begin with the two lighter inert gases.

Helium

Helium is the second lightest and the second most abundant atom in the universe. It assumes a special status in the present context as it is a very tightly bound system, having the highest first ionization energy $I_{He} = 24.6$ eV and a rather low polarizability of 0.204 Å^3.

Electron–He collision cross sections were reviewed by Zecca et al. (1996) and Khare (2001). Electron impact cross sections for ionization and excitation of He are available in the NIST database. Comprehensive calculations of the major e–He cross sections Q_T, Q_{el}, Q_{inel}, Q_{ion} and the excitation-sum ΣQ_{exc}, based on the CSP-ic method were carried out by Vinodkumar et al. (2007). The electron–atom interaction was modelled as a total complex potential V_{tot} (r, E_i), determined from the atomic charge density, obtained from (Clementi and Roetti1974, Bunge et al.1993). The non-elastic effects were contained in the imaginary term V_{abs} (r, Δ, E_i), which is based on Staszewska et al. (1984). The threshold parameter Δ in V_{abs} may be fixed at I, but a variable-Δ form of the absorption potential was also employed in the calculations (Vinodkumar et al. 2007). The variable $\Delta = \Delta(E_i)$ starts from a value somewhat lower than I, and allows for excitation channels which open up before the ionization threshold. The ionization fraction is defined through an energy dependent ratio function (as in Chapter 2)

$$R(E_i) = \frac{Q_{ion}}{Q_{inel}} \tag{3.3}$$

Turner et al. (1982) were probably the first who made a semi-empirical estimate of such a ratio for electron scattering, and predicted that ionization would be more probable than excitation above 30 eV. Turner et al. found, in the case of water molecule, that above 100 eV,

$$\frac{\sigma_{ion}}{\sigma_{ion} + \sigma_{exc}} \approx 0.75 \tag{3.4}$$

with σ_{ion} and σ_{exc} denoting ionization and electronic excitation cross sections respectively.

The CSP-ic method is developed by considering the behaviour of $R(E_i)$ at three typical energies, viz.,

(a)   the first ionization threshold I,

(b)   the peak position 'E_p' of Q_{inel}, and

(c)   a typical high energy $E_i \gg E_p$.

To establish a general theoretical procedure we write

$$R(E_i) = 0, \text{ for } E_i \le I \tag{3.5a}$$

$$= R_P, \text{ at } E_i = E_P \tag{3.5b}$$

$$= R', \text{ at } E_i \gg E_P \tag{3.5c}$$

Note that in equation (3.5c) we are more specific than in equation (2.31c) of Chapter 2. The ratio values R_p and R' (≤ 1) are obtained by invoking reasonable semi-empirical arguments. In a large number of common atomic–molecular targets, the first ionization threshold I is around 10–15 eV, with important excitation thresholds occurring at lower energies. The peak position E_p of the calculated cross section Q_{inel} takes place broadly around 50–100 eV. For a target having lower threshold I, the E_p occurs at a relatively lower energy. At an incident energy $E_i = E_p$, the cross sections for discrete transitions are reducing after reaching their peak while Q_{ion} is increasing, and therefore the unknown R_p in equation (3.5b) is expected to be between 0.5 and 1. The general trend amongst common and well-studied atoms and molecules suggests this value to be around 0.7, while it is close to 0.8 for light inert gases exhibiting higher thresholds of ionization. A notable exception in this regard is the hydrogen atom for which the said ratio is practically constant at 0.5 for intermediate and high energies (see Chapter 2). The general choice of R_p (0.8 in light inert gases and 0.7 in other targets) though reasonable, is not rigorous and is likely to introduce uncertainty in the final results. Next comes the second unknown R' occurring in equation (3.5c), which is expected to be close to unity, with the dominance of Q_{ion} at high energy. In the preliminary CSP-ic calculation, a suitable value $R' \lesssim 1$ is assumed typically at $E_i = 10E_p$. But, there is a variant of the CSP-ic, wherein R' is not assumed but generated as outlined below.

In view of the three conditions (3.5a–c) we represent $R(E_i)$ as a continuous function of a scaled variable $U = E_i/I$, as mentioned in Chapter 2. Thus,

$$R(E_i) = 1 - C_1 \left[\frac{C_2}{U + a} + \frac{\ln U}{U} \right] \tag{3.6}$$

The first condition (3.5a) is exact, while the second one (3.5b) requires a semi-empirical input $R_p = 0.8$ for helium. The next unknown R' is determined by starting initially with parameter $a = 0$ in equation (3.6) and employing two conditions (3.5a and 3.5b) to find the initial values of parameters C_1 and C_2. Plugging these values into the two-parameter version of equation (3.6) we obtain a preliminary estimate of Q_{ion} through our calculated Q_{inel}. This provides us with an estimated R', which is then used as the input in equation (3.5c). The next step is to go back to the full 3-parameter expression (3.6) and obtain the desired ionization cross section. Thus in the variant CSP-ic method the above procedure is carried out iteratively till consistency is achieved in the values of the three parameters for a given target. The crucial input is R_p for which an assessment is possible for the standard target He. Table 3.1 exhibits the final values of the CSP-ic parameters R_p, R', a, C_1 and C_2. Here, the parameter R' is obtained at a high energy, viz., $E_i = 10$. $E_P = 1000$ eV.

Table 3.1   Dimensionless CSP-ic parameters for e-He scattering

R_p at $E_P = 100$ eV	R' at 1000 eV	α	C_1	C_2
0.799	0.9795	8.029	−1.584	−5.70

Let us have a look at the theoretical e–He results derived in CSP-ic (Vinodkumar et al. 2007) in the energy range, starting from the ionization threshold to 2000 eV, as shown Figure 3.1. Theoretical cross sections Q_T have been compared in this figure with three experimental data sets (Baek and Grosswendt 2003, Blaauw et al. 1980, Nickel et al. 1985) and theoretical estimates of de Heer et al. (1977). It should be noted that two of these data sets are more than two decades old. The Q_T of e–He system determined in the CSP-ic approximation is depicted by the top most curve (Figure 3.1), and agree well with experimental data above 20 eV. At 1000 eV, the CSP-ic Q_T is 0.18 Å^2 as against 0.22 Å^2 derived using high energy EBS theory (Khare 2001). The calculated elastic cross sections Q_{el}, shown by the middle curve in this figure, merge with Q_T at the lower end of energies as expected and agree with the earlier data reported in Zecca et al. (1996). The lower curve in Figure 3.1 shows the ionization cross sections Q_{ion} calculated using the CSP-ic is in good agreement with the experimental data of Krishnakumar and Srivastava (1988) and the calculations of Kim and co-workers using the Binary Encounter Bethe or **(BEB)** theory (see NIST database). The calculated Q_{ion}, though cumulative, are dominated by single ionization cross sections of He, since contributions from double ionization and inner shell processes are small. The dependence of different TCS as functions of incident energy makes an interesting study. The cumulative excitation cross sections ΣQ_{exc} are not shown in the Figure 3.1, but the calculations (Vinodkumar et al. 2007) showed that ΣQ_{exc}, Q_{inel} and Q_{ion} exhibited broad peaks around 90, 100 and 120 eV respectively, and this is in accordance with the corresponding threshold energies of the helium atom. It should be noted that the maximum Q_{ion} is less than the static geometrical area $\pi <r>^2$ with $<r> = 0.49$ Å for He atom.

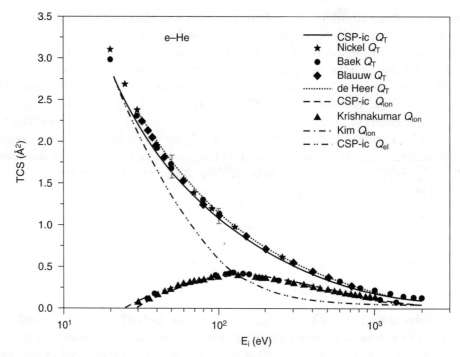

Figure 3.1   Various TCS of e–He scattering (Vinodkumar et al. 2007), along with compared data

In order to further assess the validity of the CSP-ic method, the cross sections Q_{ion}, ΣQ_{exc}, Q_{inel} and the ratio $R_p = Q_{ion}/Q_{inel}$ were drawn from NIST database and compared with CSP-ic results. Table 3.2 compares these cross sections at 100 eV energy corresponding to the peak of the inelastic cross section. Good agreement is found for Q_{ion}, ΣQ_{exc} and Q_{inel}, at least within the known uncertainties. The input value $R_p = 0.799$ assumed in the theory is in very good agreement with the corresponding value 0.787, derived from the NIST database, and the small difference in R_p does not appreciably affect the values of the final cross sections derived using CSP-ic method. Note that validation of the input R_p is not possible in most targets, and this necessitates the need for a semi-empirical assumption of R_p. At the Q_{ion} peak (120 eV), Q_{el} and Q_{inel} are almost equal and Q_{ion} and ΣQ_{exc} are 42% and 7% respectively of the total (complete) cross section Q_T.

Table 3.2   Comparison of e–He TCS (in Å^2) at the inelastic peak $E_p = 100$ eV, (a) Vinodkumar et al. (2007); (b) data compiled from NIST database

Q_T	Q_{el}	Q_{inel}		Q_{ion}		ΣQ_{exc}		$R_p = Q_{ion}/Q_{inel}$	
(a)	*(a)*	*(a)*	*(b)*	*(a)*	*(b)*	*(a)*	*(b)*	*(a)*	*(b)*
1.11	0.60	0.51	0.47	0.41	0.37	0.10	0.095	0.80	0.787

Neon

Electron scattering cross sections with Ne, Ar, Kr and Xe, were reviewed some years ago (Zecca et al. 1996, Khare 2001) while more recently Vinodkumar et al. (2007) calculated the electron scattering TCS for these atoms from threshold to 2000 eV employing CSP-ic.

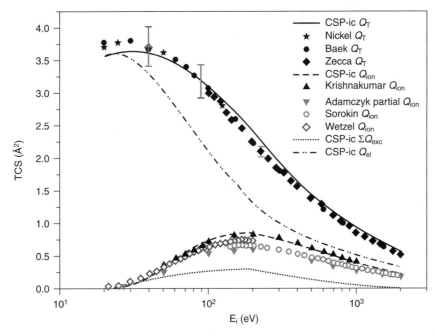

Figure 3.2 Various TCS of *e*–Ne scattering (Vinodkumar et al. 2007) along with compared data.

Figure 3.2 presents the Q_T, Q_{el} and Q_{ion} of neon. The CSP-ic Q_T, shown by the upper most curve in this figure, are in a very good agreement with available experimental results (Nickel et al. 1985, Zecca et al. 1996, Baek and Grosswendt 2003) above 30 eV. The Q_{el} (dash double dot curve) calculated using CSP-ic is slightly lower than the values reported by Zecca et al. (1996). Both Q_T and Q_{el} curves exhibit a gentle broad maximum in the region of 20 eV. One can see from this figure that elastic scattering is appreciable even at high energies.

The Q_{ion} values derived using CSP-ic are compared with four different experiments conducted by Wetzel et al. (1987), Krishnakumar and Srivastava (1988), Sorokin et al. (1998) and a rather old measurement by Adamczyk et al. (1966). All of the experimental data agree within 10–12% but the CSP-ic Q_{ion} are somewhat higher than experiment, suggesting that the input value $R_p = 0.76$ may be a little high. The lowest curve in Figure 3.2 presents a summed total excitation cross section ΣQ_{exc}, which shows a broad maximum in the range of 100–200 eV, before falling off with energy. It should be noted that the ΣQ_{exc} is an upper limit but this estimate is important since absolute cross section data on electron impact excitations are rather scarce in inert gases and other targets.

3.2.2 Argon, Krypton and Xenon

Argon

Consider next the different cross sections for electron argon collisions depicted in Figure 3.3 (Vinodkumar et al. 2007). Excellent agreement is found between the Q_T results derived using the CSP-ic method and the four sets of experimental data (Nickel et al. 1985, Zecca et al. 1996, Szmytkowski et al. 1996, Baek and Grosswendt 2003), while the SCOP derived Q_T data of Jain et al. (1990) and Karim and Jain (1989) are found to underestimate the cross section at higher energies. However CSP-ic Q_{el} are close to those of Jain et al. (1990). In the lower curves CSP-ic Q_{ion} values are compared with the experimental results of Nagy et al. (1980), Krishnakumar and Srivastava (1988), Sorokin et al. (2000). The measured data of Krishnakumar and Srivastava are higher than the theoretical values below about 100 eV. At 100 eV the CSP-ic values for argon are

$$Q_T = 7.45 \text{ Å}^2, Q_{el} = 4.10 \text{ Å}^2, Q_{ion} = 2.54 \text{ Å}^2 \text{ and } \Sigma Q_{exc} = 0.81 \text{ Å}^2$$

It is interesting to determine the relative importance of different TCS for Ar. Figure 3.4 adopted from Vinodkumar et al. (2007) shows the five cross sections for electron scattering from argon at the energy where Q_{ion} is maximum i.e., 125 eV. At this energy Q_{el} is, within known uncertainties, equal to Q_{inel}, and Q_{ion} is nearly three times ΣQ_{exc}.

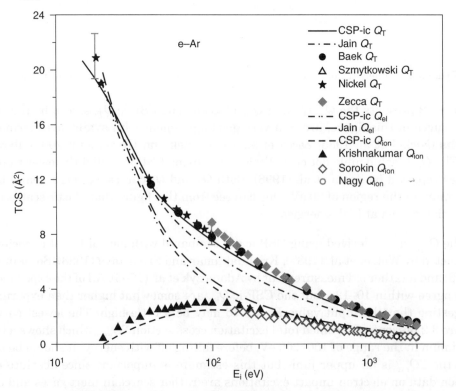

Figure 3.3   TCS for *e*–Ar scattering (Vinodkumar et al. 2007)

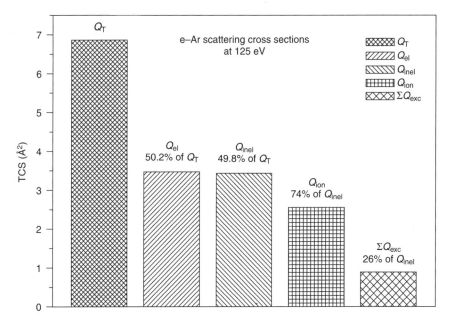

Figure 3.4 Bar-chart showing the relative contribution of various scattering processes with argon near the ionization peak, 125 eV (Vinodkumar et al. 2007)

Krypton, Xenon and Radon

Turning now to the heavier inert-gas atoms Kr and Xe, we must note that relativistic (Dirac–Hartree–Fock–Slater) charge-density expressions were employed in the calculations of Vinodkumar et al. (2007) to construct the total complex potential. The TCS of krypton atom are presented in Figure 3.5 with Q_T (upper curves) and Q_{ion} (lower curves). The CSP-ic Q_T cross sections Vinodkumar et al. (2007) agree well with the measured data assembled by Zecca et al. (1996) as well as Dababneh et al. (1982) but the theoretical values of Jain et al. (1990) are once again lower at higher energies. Electron scattering from krypton has a high Q_{el}, not shown in figure, and the CSP-ic values are consistent with those tabulated in Zecca et al. (1996). At the ionization peak i.e., at 70 eV, the CSP-ic results for krypton are as follows.

$$Q_T = 12.13 \text{ Å}^2, \ Q_{el} = 7.27 \text{ Å}^2 \text{ and } Q_{ion} = 4.20 \text{ Å}^2$$

Moreover there is a good agreement between the Q_{ion} results from the CSP-ic calculations and the measured data of Nagy et al. (1980), Krishnakumar and Srivastava (1988) and Sorokin et al. (2000).

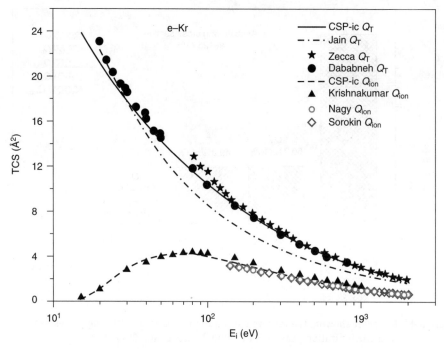

Figure 3.5   TCS for *e*–Kr scattering (Vinodkumar et al. 2007)

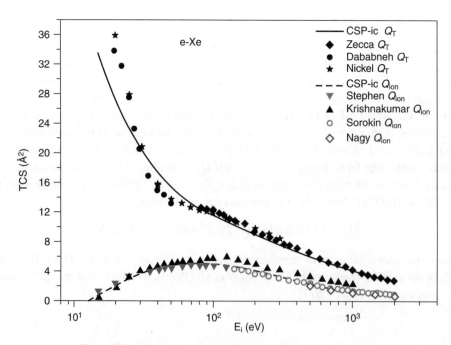

Figure 3.6   TCS for *e*-Xe scattering (Vinodkumar et al. 2007)

Figure 3.6 presents TCS for e–xenon scattering. Comparison of the CSP-ic values Vinodkumar et al. (2007) of Q_T with Zecca et al. (1996), Nickel et al. (1985) and Dababneh et al. (1980, 1982) indicate that CSP-ic values tend to lie below the measurements at lower energies and do not replicate the small minimum recorded in the experiments around 50 eV. Elastic cross sections Q_{el} for this target, not shown in this figure, are once again high. There is good agreement between experimental Q_{ion} cross sections (Nagy et al. 1980, Stephen and Maerk 1984, Krishnakumar and Srivastava 1988, Sorokin et al. 2000) and the CSP-ic results of Vinodkumar et al. (2007). In the case of Xe, the ionization peak occurs at 60 eV, and at that energy

$$Q_T = 13.03 \text{ Å}^2, Q_{el} = 6.14 \text{ Å}^2 \text{ and } Q_{ion} = 5.36 \text{ Å}^2.$$

Electron scattering from the last member of the inert-gas family, Radon (Rn), has not been investigated experimentally, perhaps due to its radioactive nature. The CSP-ic method was employed by Joshi et al. (2016) to study the electron scattering with the Radon atom under simplified assumptions. In this case, the calculated peak Q_{ion} occurred at 55 eV although the cross section, 4.80 Å^2 appears to be rather low.

Concluding remarks on the inert gases

In concluding the discussion on inert-gas targets, consider a few general trends. Table 3.3 presents several features of electron scattering from the inert gas targets. The three lighter members of this group are characterized by relatively large first ionization energies. The ratios of both Q_{el} and Q_{inel}, found to be around 50% of the total cross section Q_T at E_p (peak of Q_{inel}), are remarkably similar.

Table 3.3 A comparative study of theoretical results (Vinodkumar et al. 2007) on inert-gas atoms, at their respective peaks of Q_{inel}

Target atom	First Ionization Energy eV	% of Q_T at respective E_P	
		Q_{el}	Q_{inel}
He	24.6	54.0	46.0
Ne	21.6	54.4	45.6
Ar	15.6	50.2	49.8
Kr	14.0	58.4	41.6
Xe	12.1	52.1	47.9

Given the similarities in the electron scattering from the inert gases, it is tempting to represent the energy dependence of the cross sections Q_T of the inert gas atoms by an analytical formula. Vinodkumar et al. (2007) considered an analytical formula for the total cross sections of the inert gases above 500 eV, as follows

$$Q_T(E_i) = A\,(\alpha_d/E_i)^B \tag{3.7}$$

where 'A' and 'B' are the target-dependent parameters and α_d is the atomic static electric dipole polarizability. Table 3.4 shows the numerical values of their fitting parameters for He, Ne, Ar, Kr and Xe. The index parameter 'B' is around 0.7 for all inert gases except He for which it is 0.9. The same dependence of Q_T on E_i was observed earlier for various 18-electron systems, namely Ar, H_2S, CH_3OH, CH_3NH_2, CH_3F and C_2H_6 (Joshipura and Vinodkumar 1999).

Table 3.4   Fitting parameters A and B of the formula (3.7) (see text)

Atomic target	A	B
He	306.5	0.90
Ne	200.6	0.69
Ar	165.6	0.69
Kr	127.5	0.61
Xe	120.0	0.59

The dependence of Q_T on the polarizability is also given approximately by $(\alpha_d)^{0.7}$ as observed earlier by Szmytkowski and Krysztofowicz (1995).

The analytical representation expressed in equation (3.7) is not unique and several versions have been attempted. What does a formula like this convey? A physical interpretation is possible by considering a square-root dependence of the cross section Q_T on incident energy i.e.,

$$Q_T \alpha \ (E_i)^{-0.5} \tag{3.8}$$

This simply means that the cross section is inversely proportional to the speed of the incident electron, making the scattering probability proportional to time spent by the electron in the vicinity of the target region.

Inert gases are the best test systems for theoretical as well as experimental investigations, and hence they can act as experimental calibrants for other atomic and molecular systems. CSP-ic theoretical results for various total cross sections, in general, show good agreement with other theoretical and experimental investigations. Moreover, a general picture of the various processes involved is obtained through the relative contribution of various cross sections to the total (complete) cross section.

3.3   OXYGEN, NITROGEN, AND CARBON

In terms of relative abundance in the universe, the elements oxygen, nitrogen and carbon (along with neon) are next to H and He; and O, N and C form a large number of the common and light molecules. In this section we discuss TCS for these three elements. For all three atoms experimental data are scarce, except for ionization, since it is hard to create high purity beams of such atoms.

3.3.1 Oxygen

Interactions of oxygen atoms with electrons play an important role in atmospheric and planetary science but our knowledge of the relevant cross sections is poor. Reviews of electron scattering from atomic oxygen were collated by Zecca et al. (1996) and Itikawa and Ichimura (1990). Joshipura and Patel (1993) calculated differential as well as total elastic cross sections employing model potentials, while Joshipura et al. (2006) used CSP-ic, along the lines of equations (3.5)-(3.6), to derive Q_T at intermediate and high energies. For the O atom the values of the three parameters in the expression (3.6) were found to be as follows.

$$a = 6.92, C_1 = -1.12 \text{ and } C_2 = -7.07$$

The peak position (E_p) of the inelastic cross section Q_{inel} was found to occur at 70 eV, while the peak of the Q_{ion}, denoted by ε_{ion}, occurred at 100 eV. At E_p the semi-empirical input value of R_p was employed by Joshipura et al. (2006) to be 0.71 and the ratio $R(E_i)$ increases to 0.75 at ε_{ion}= 100 eV. CSP-ic calculations at 100 eV, yielded Q_T = 5.4 Å² compared to the two experimental values of 4.55 Å² and 5.9 Å² quoted in Zecca et al. (1996). For an indirect but a better comparison, the independent atom model (from Chapter 2) may be used and we find that at 1000 eV, the ratio of atomic to molecular oxygen TCS $Q_T(O)/Q_T(O_2)$ is close to 0.5, as expected. We shall return briefly to the O atom in comparison with the O_2 molecule in the next chapter. At high energies, above 1000 eV or more, the elastic TCS Q_{el} of atomic oxygen (Joshipura et al. 2006) tends to be closer to Q_{ion}.

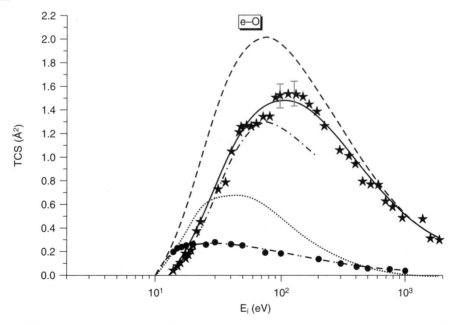

Figure 3.7 Cross sections of *e–O* collisions (Joshipura et al. 2006); *topmost curve:* calculated Q_{inel} by the CSP-*ic* method; Q_{ion} *continuous curve:* measured Q_{ion} data (Thompson et al. 1995); ★: Q_{ion} DM formula (Deutsch et al. 2001); — · — · —; *the lowest two curves:* calculated ΣQ_{exc}. the excitation-sum from two data-sets — ·· — and • by Johnson et al. (2003)

Figure 3.7 presents three cross sections Q_{inel}, Q_{ion} and ΣQ_{exc}. The theoretical CSP-ic TCS Q_{inel} is plotted as the topmost curve but has no direct experimental comparison. However CSP-ic Q_{ion} are in good agreement with the experimental data measured by Thompson et al. (1995). The measurements are quoted with a 7% error at the peak and, although the oxygen beam may contain a small amount of metastable states of O, their effect on the measured cross section is believed to be small. The question of metastable contamination in connection with other atomic beam targets, will be taken up later in this chapter.

Ionization calculations using the Deutsch–Maerk or DM formula (Deutsch et al 2001) are also included in Figure 3.7 and although limited to 200 eV the DM values are found to be lower in magnitude. Finally at 100 eV, the BEB theory (NIST database) predicts the value of Q_{ion} to be 1.24 Å^2 as against the CSP-ic value 1.50 Å^2. Thus the CSP-ic data appears to most closely match the single experimental measurement of Q_{ion}.

Once again CSP-ic calculations seek to determine the relative importance of excitation as against ionization in the form of the excitation sum ΣQ_{exc}. Qualitative agreement, especially in the peak position, is found with Zecca et al. (1996). Johnson et al. (2003) had investigated electron-induced excitation-emissions from atomic O by considering four important transitions in the VUV region. We have included in Figure 3.7 (reproduced from Joshipura et al. 2006), the sum total of these experimental total cross sections (within 24% error). Johnson et al. (2003) have also given model calculations for the excitations of the said four states of O (see Figure 3.7). The difference between the CSP-ic excitation–sum and the summed data of Johnson et al., can be attributed partly to the low-lying states not being included in the sum and partly to approximations made in the CSP-ic theory. The difference narrows at energies beyond the ionization peak, as one can expect.

Towards lower energies, typically below the ionization threshold, the CSP-ic is not reliable in view of the failure of the model potentials employed. At intermediate to high energies, when the methodology is reliable, a few important conclusions can be drawn. Thus, one finds (Figure 3.7) that at about 25 eV the cumulative excitation and ionization cross sections for atomic O have an equal share in the Q_{inel}. This can be considered as yet another energy 'land-mark', and this is found to be the case with practically all targets having moderate ionization thresholds in the range of about 12–15 eV. At high impact energies above 1000 eV, Q_{inel} and Q_{ion} are hardly distinguishable.

To analyze these theoretical results further, Figure 3.8 from Joshipura et al., shows the relative contributions of different e–O collision processes at the peak of Q_{ion} 100 eV.

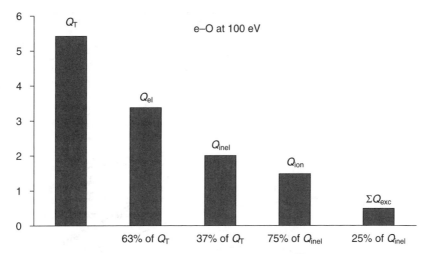

Figure 3.8 A bar-chart of relative contributions of 100 eV TCS (Å^2) for atomic oxygen (Joshipura et al. 2006)

3.3.2 Nitrogen

Our next target is atomic nitrogen which is also important in the study of the Earth's ionosphere as well as the upper atmospheres of several other planets. There appear to be neither review papers nor any recommended data sets on electron–nitrogen atomic collisions. Theoretical calculations using CSP-ic with V_{abs} in variable-Δ form were reported by Joshipura et al. (2009). Figure 3.9 depicts the different total cross sections for e–N collisions. Once again the top-most curve represents Q_T. In the absence of any experimental data on Q_T for the N-atom, the independent-atom-model (IAM) allows a comparison of calculated Q_T (N) with $Q_T(N_2)/2$, where Q_T (N$_2$) were obtained from the recommended data on N$_2$ molecule (Itikawa 2006), to be discussed in the next chapter. Agreement is found to be satisfactory above 100 eV.

Also shown in Figure 3.9 are the calculated Q_{el}, merging with Q_T at lower energies, and lying close to Q_{ion} at high energies. The CSP-ic calculations of Q_{el} may be compared with results of two of the old investigations (Blaha and Davis 1975, Ormonde et al. 1973) but there is little or no agreement mainly due to differences in the potential models.

The nitrogen atom with its first ionization threshold at 14.53 eV, exhibits a peak in the ionization cross section near 100 eV. In Figure 3.10, the ionization and excitation-sum results of CSP-ic are shown along the BEB values of Kim and Desclaux (2002) and the IAM value Q_{ion} (N$_2$)/2 obtained from Itikawa (2006).The BEB values are once again lower than the results of Joshipura et al. (2009) but in good agreement with the IAM value Q_{ion} (N$_2$)/2 above 300 eV.

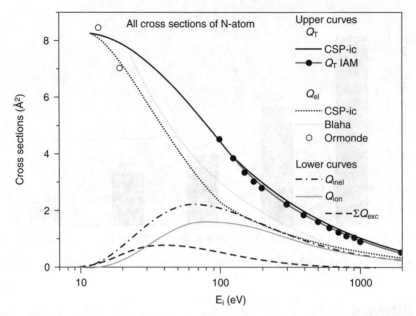

Figure 3.9 Various total cross sections of e–N atom collisions (Joshipura et al. 2009)

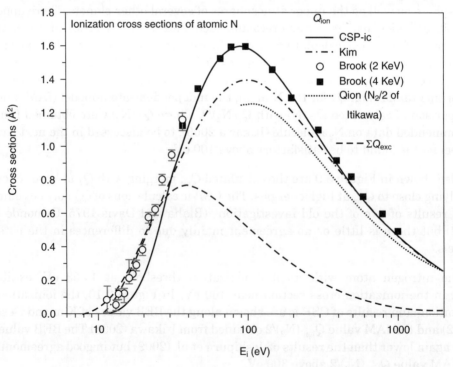

Figure 3.10 Ionization and cumulative excitation in e–N scattering (Joshipura et al. 2009); Cross section Q_{ion}; continuous curve CSP-ic, dash dot BEB of Kim and Desclaux (2002), measured data; circle and box – two data sets by Brook et al. (1978), Q_{ion} (N$_2$/2) IAM (see text), lowest curve ΣQ_{exc} CSP-ic

CSP-ic results can also be compared with the measured data of Brook et al. (1978) who had formed neutral N atom beams by charge exchange with N ions at two different ion energies (2 keV and 4 keV) of the atomic beam, hence the two sets of experimental data in Figure 3.10. Very good agreement is seen at intermediate and high energies, but below 30 eV the experimental ionization data gradually tend to lie well above the present theoretical curve. However, a closer look at the measured data of Brook et al. (1978) shows small but finite Q_{ion} at and below the ionization threshold I = 14.53 eV. This particular behaviour is an indication of the presence of metastable excited species in the experimental atomic N beam. The ionization signals at or below the threshold (I) of the ground-state N atom arise due to the presence of excited metastable atoms having lower ionization thresholds. Therefore it is to be expected that CSP-ic values should lie lower than the Brook et al. data since our 'theoretical' target consists of N atoms purely in the ground state. Electron ionization out of the initially excited metastable N atoms is addressed separately in a later section of this chapter.

3.3.3 Carbon

Atomic carbon is found to exist in various forms in the Sun and in stars, planets, comets and such, but electron scattering from the open shell carbon atom has, to date, not received much attention.

The CSP-ic method has been used by Pandya (2013), Pandya and Joshipura (2014) to examine e–carbon atomic scattering, and the dimensionless three parameters in this case are found to be a = 8.342, C_1 = –1.283 and C_2 = –7.279. The first ionization energy of carbon is I = 11.26 eV and accordingly the inelastic peak position E_p is found to occur around 60 eV. The CSP-ic results are shown in Figure 3.11 (Pandya 2013). There appear to be no measured data for Q_T or Q_{el} for atomic C, while approximate values of $(Q_{el} + Q_{inel})$ for the carbon atom were estimated by Joshipura and Patel (1994). The Q_{el} results (Pandya 2013) show the expected behaviour and tend to be nearly equal to Q_{ion} above 300 eV. The CSP-ic cross sections Q_{ion} displayed in Figure 3.11 are in good agreement with the experiments of Brook et al. (1978), and the BEB values of Kim and Desclaux (2002). Bartlet and Stelbovics (2004) have calculated total ionization cross sections of atoms up to Z = 54, using a plane wave Born approximation. For the carbon atom the theoretical data of Bartlet and Stelbovics agrees with the CSP-ic values above 300 eV but are somewhat lower at lower energies. Amongst the three targets O, N and C examined in this subsection, the carbon atom exhibits the highest *peak* cross section Q_{ion} = 2.22 Å^2 and that at comparatively lowest incident energy of 60 eV.

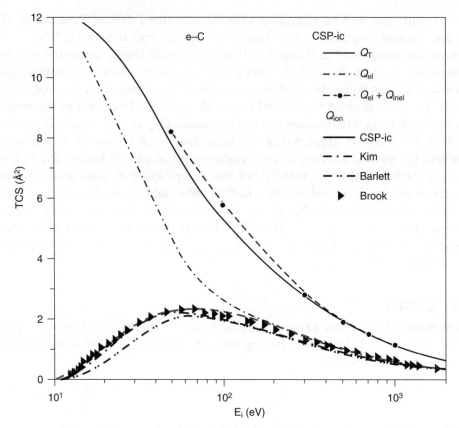

Figure 3.11   TCS of e–C scattering (Pandya 2013) along with comparisons; from top: - • - (Q_{el} + Q_{inel}) Joshipura and Patel (1994), next two curves CSP-ic, ——— Q_T and - - - Q_{el}; lower four curves, Q_{ion} as indicated inside

3.4   OTHER ATOMIC TARGETS

In this section results are presented and discussed on selected atoms, while passing remarks are given on a few others.

3.4.1   Beryllium and Boron

Collisions of electrons involving beryllium atoms are important since Be has been chosen as the material for the walls of the International Thermonuclear Experimental Reactor – ITER. A literature review does not indicate any experimental data but several theoretical results are available including two CSP-ic approximations, namely Chaudhari et al. (2015), and Joshi et al. (2016). Figure 3.12 displays the energy dependence of the calculated TCS of e–Be collisions (Chaudhari et al. 2015). The total cross sections Q_T are quite large compared to the previous light atoms, especially at lower energies and the contribution from elastic scattering is seen to be fairly large.

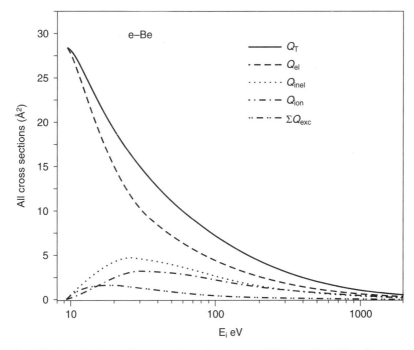

Figure 3.12   TCS plot of e–Be collisions evaluated using the CSP-ic method (Chaudhari et al. 2015)

Different estimates of the Q_{ion} for beryllium atoms are shown in Figure 3.13. Results include the Deutsch–Maerk model (DM), BEB calculations (Maihom et al. 2013) and the R-matrix with pseudo-states (RMPS) calculation by Zakrzewski and Ortiz (1996). The CCC results are from the data paper of Bartlett and Stelbovics (2004). Other calculated data included here are the DW (distorted wave) calculations (Colgan et al. 2003) and of Fursa and Bray (1997). There is little agreement between the different models. The CSP-ic Q_{ion} shown by the black continuous curve falls almost in the middle of the six sets data shown in Figure 3.13. The DM values show the largest TCS at the peak, followed by the BEB results. The other two theoretical results, that is, CCC and Fursa and Bray (1997) are lower. A reasonably high peak cross section, Q_{ion} (2.5 Å^2 around 40 eV) as calculated by CSP-ic is in line with its atomic properties as against other atoms.

In the case of electron scattering from boron we restrict our discussion to ionization results shown in Figure 3.14. Here, we again compare Q_{ion} calculated using different methods including the DM formula (Margreiter et al. 1994, Maihom et al. 2013), BEB results (Kim and Stone 2001) and recommended data of Moores (1996) and also the rather old data from Lotz (1970) and Stingl (1972). The CSP-ic calculations by Chaudhari et al. (2015) are once again in the middle of the range of presented values with the maximum values obtained in the DM and BEB formalisms. Q_{ion} of boron is 3.5 Å^2 at about 40 eV. For electron impact ionization of atomic Be and B, the CSP-ic methodology was also employed by Goswami et al. (2013), with results and comparisons similar to those of Figures 3.13 and 3.14.

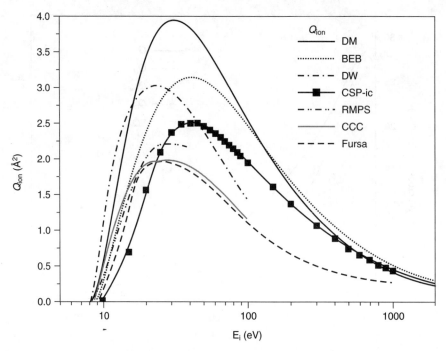

Figure 3.13   Electron collisional ionization of Be atoms; continuous curve CSP-ic by Chaudhari et al. (2015) and other results

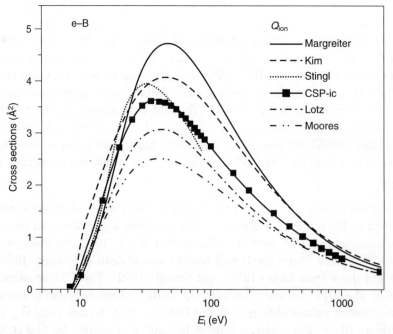

Figure 3.14   Electron collisional ionization of Boron atoms, CSP-ic by Chaudhari et al. (2015) and other results

This is the appropriate stage for considering the electron scattering from alkali atoms. These may be considered exotic in view of their low excitation and ionization thresholds, large sizes and very high polarizabilities. Earlier data sets in this regard were compiled by Zecca et al. (1996). In these targets, the important excitation thresholds also occur at very low energies, and dipole allowed transitions dominate over ionization even at relatively high energies. In Figure 3.15, we have compared two of the available electron impact cross sections of Li atoms. Here the excitation cross section Q_{exc} (2s → 2p) as calculated in the 'BEf-scaled' method is adopted from the NIST database, while the Q_{ion} result (lower curve) is that of BEB calculations given by Irikura[i]. No CSP-ic result is presented here, but it becomes clear from the figure that excitation dominates over ionization in e–Li scattering.

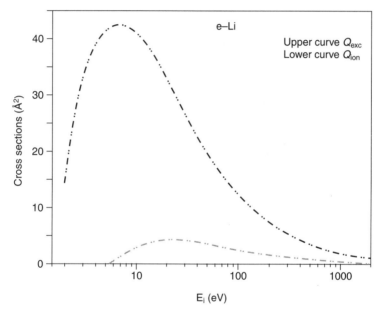

Figure 3.15 Electron scattering with the Lithium atom; Upper curve Q_{exc} (2s → 2p) BEf-scaled excitation cross section (NIST database), Lower curve Q_{ion} BEB ionization data (Irikura 2017)[ii]

3.4.2 Halogen atoms

Collisions of electrons with atomic halogens have received considerable attention in view of their role in stratospheric ozone depletion. Due to their high reactivity and difficulty in preparing atomic beams of halogens there are few experiments and therefore scattering data must be largely derived from theory. Electron ionization of halogen atoms F, Cl, Br and I together with their homonuclear diatomic molecules, was investigated in a simpler version of the CSP-ic theory more than a decade ago by Joshipura and Limbachiya (2002). These calculations were revised by Vinodkumar et al. (2010).

---

i, ii   Irikura Karl K. 2017. Private communication (unpublished).

With $I = 17.42$ eV and $\alpha_d = 0.56$ Å^3 atomic fluorine is a weak scatterer. Figure 3.16 shows the CSP-ic TCS for electron induced ionization of the fluorine atom as compared with measurements of Hayes et al. (1987) and the DM formula results of Margreiter et al (1994). The measured data of Hayes et al. (1987) albeit with large error bars tend to be ~ 10% lower than the theoretical values in the peak region. The DM calculations are lower than the CSP-ic results and agree better with the latest close-coupling calculations of Gedeon et al. (2014) for single ionization of F atoms by incident electrons, not shown in Figure 3.16.

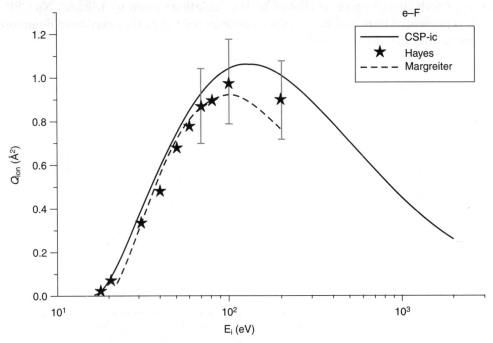

Figure 3.16 Total ionization cross sections of Flourine atoms; continuous curve from Joshipura and Limbachiya (2002); ★ experimental data from Hayes et al. (1987), DM formula from Margreiter et al. (1994)

In the work of Gedeon et al. (2014), devoted to several atomic targets, general aspects on the uncertainty estimates in the theoretical electron–atom collision data are also discussed; ascribing uncertainties to theoretical data is a relatively new concept but one that is increasingly necessary if theoretical data is to be the major input into models and simulations (Chung et al. 2016). Uncertainty in theoretical evaluation of collision cross sections depends on the kind of process being investigated, and it involves errors arising from target representation including the accuracy of the structure properties used as inputs, and the nature of theoretical method adopted. In the CSP-ic methodology accurately known basic target properties are used and therefore the sources of uncertainty are the model potentials employed and the assumptions made in this methodology. However, in most of the cases examined in this book

CSP-ic ionization cross sections are within experimental errors and may be estimated at about 10–15%.

Ionization cross sections for the other halogen atoms along with comparisons with other available data were presented by Vinodkumar et al. (2010) in their paper.

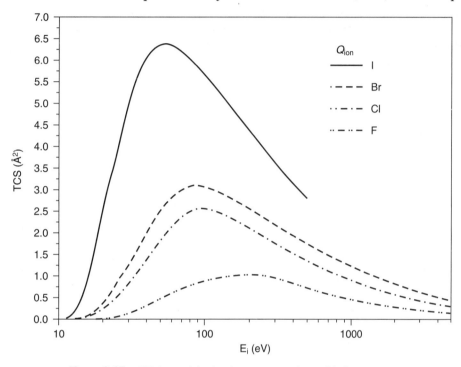

Figure 3.17 CSP-ic total ionization cross sections of halogen atoms
I, Br, Cl, and F from top to bottom

In Figure 3.17, the calculated Q_{ion} for F, Cl, Br and I targets based on the CSP-ic data are shown (Joshipura and Limbachiya 2002). The relative shapes and magnitudes of the Q_{ion}, and the peak positions are in accordance with the basic atomic properties, namely; (i) a shift in ε_{ion} the maximum of the Q_{ion} cross section to lower energies as the ionization energy of the target is lowered and (ii) a larger cross section for heavier targets in accordance with the increasing geometric size of the target.

3.4.3 Silicon, Phosphorus, Sulphur, and Germanium

Several theoretical estimates of the Q_{ion} of silicon atoms have been reported including the DM calculations of Margreiter et al. (1994), the first Born calculations of Bartlett and Stelbovics (2004) and the CSP-ic results (with variable-Δ form in the absorption potential V_{abs}) of Gangopadhyay (2008). These may be compared with the single set of experimental data of Freund et al. (1990) shown in Figure 3.18. None of the theoretical methods adequately match the experimental data which peaks at lower energy, though all the theoretical methods agree on the shape and position of the peak which Gangopadhyay estimates at 6.9 Å² around 40 eV. Figure 3.18 also shows the only evaluation of the summed excitation cross section for silicon atom (Gangopadhyay 2008).

Electron impact ionization with phosphorus and the other atoms of periodic group-VA has been investigated by Joshipura et al. (2009) using the CSP-ic approach and the Q_{ion} were found to be much lower than the measured data (Freund et al. 1990). Electron impact ionization of allotropes of phosphorus and arsenic were investigated by Bhutadia et al. (2015) using the CSP-ic while Kaur et al. (2015) have employed CSP-ic to calculate Q_{ion} of sulphur atoms.

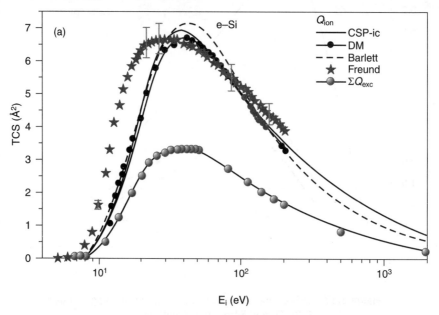

Figure 3.18 Electron ionization of Si atoms; Q_{ion}: CSP-ic: *continuous curve*: Gangopadhyay (2008); *dots*: DM (Margreiter et al. 1994); *dashes*: Bartlett and Stelbovics (2004); *stars*: experimental data Freund et al. (1990); ΣQ_{exc}: *lowest curve*: CSP-ic

It is of interest to compare the theoretical Q_{ion} for a set of three targets carbon, silicon and germanium, the first three members of the periodic group IV. Ionization calculations based on CSP-ic with V_{abs} in the fixed-Δ model (Patel 2017) are shown in Figure 3.19. The results are again in accordance with the respective atomic properties.

Before moving on to other atomic targets, it is important to examine, in some detail, the elaborate experimental work on electron impact ionization of a large number of atoms by Freund et al. (1990). They measured, with the same experimental apparatus, absolute total ionization cross sections from 0 to 200 eV, for single ionization of 16 atoms viz., Mg, Fe, Cu, Ag, Al, Si, Ge, Sn, Pb, P, As, Sb, Bi, S, Se, and Te, at 10% uncertainty. These comprehensive measurements offer a very good data set for comparison over a wide range of atomic targets. Double and triple ionization TCS of some atomic species were also reported. They observed that single-ionization peak Q_{ion} decreased monotonically across the periodic table rows from (Al, Ga, In) to (Ar, Ke, Xe).

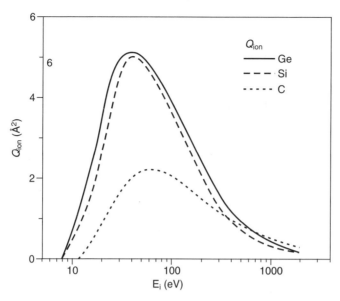

Figure 3.19 Q_{ion} for C, Si, and Ge atomic targets calculated in CSP-ic (Patel 2017)

Freund et al. discussed whether metastable states could have been populated in the process of the formation of the neutral atomic beams used for scattering experiments. If there is a sizeable component of metastable states in the target beam then the theoretical ground-state results are expected to be lower than the experimental data, see Kaur et al. (2015). Santos and Parente (2008) using the BEB method made an attempt to determine a suitable admixture of the ground and the excited state(s) in the experimental atomic beams that replicated the experimental data of group-VA atoms P, As, Sb and Bi; their results suggest that the experiment of Freund et al. did indeed have a sizeable concentration of metastable states. We will discuss electron scattering from metastable states further, later, in this chapter.

3.4.4   Aluminum and Copper

When the first ionization energy of the target is rather low, typically less than 10 eV, the magnitude of Q_{ion} is larger and the maximum of the cross section is at lower incident energy than when the target has a first ionization energy above 10 eV. This is clearly seen in electron scattering with metal-vapour atoms such as Aℓ and Cu. Electron impact ionization of aluminium (vapour) atoms has been reported by Deutsch et al. (2001) using the DM method, Kim and Stone (2001) using the BEB approximation, Bartlett and Stelbovics (2002) using first Born approximation, and Joshipura et al. (2006) using CSP-ic. These different theoretical calculations are compared with the experimental data of Freund et al. (1990) in Figure 3.20. The low ionization threshold I = 5.99 eV for Al atom results in a peak ε_{ion} in the Q_{ion} between 20 and 30 eV. However, there are significant differences between the various theories. Q_{inel} and ΣQ_{exc} derived using the CSP-ic method, are also shown in the figure. Both these cross sections are found to peak at very low energies.

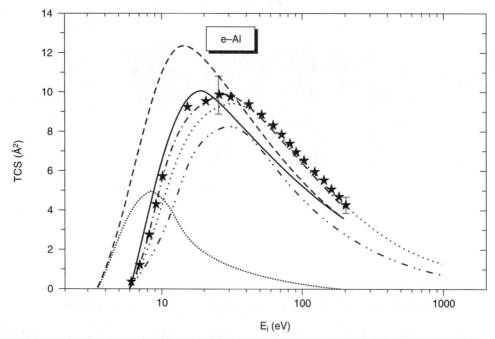

Figure 3.20   Electron scattering from Aluminium (Joshipura et al. 2006); *top-most curve*: Q_{inel}; *continuous curve*: Q_{ion}, by CSP-ic; *dash-dot curve*: Q_{ion}, by DM (Deutsch et al. 2001); *dotted curve*: Q_{ion} by BEB (Kim and Stone 2001); *lower chain curve*: Q_{ion} by Bartlett and Stelbovics (2002); ★: Q_{ion}, measured data (Freund et al. 1990); *the lowest curve*: ΣQ_{exc} by CSP-ic

The CSP-ic Q_{ion} cross sections agree with the experimental data of Freund et al. (1990) up to about 20 eV, beyond which the experimental data are higher and lie closer to the derived Q_{inel}. CSP-ic and DM results of Deutsch et al. (2001) are in broad agreement but the accurate first-Born results of Bartlett and Stelbovics (2002) are consistently lower. The results of Kim and Stone (2001) are different in shape from those of CSP-ic. However, it should be noted that the Kim and Stone data includes auto-ionization contributions and a small effect of multiple ionizations, these authors evaluating the excitation–auto-ionization contribution using the scaled Born approximation and adding these to a simple BEB derivation to obtain their final displayed Q_{ion}.

Similar observations on the cross sections can be made when studying electron impact ionization of copper, shown in Figure 3.21. The first ionization energy of Cu atom is $I = 7.73$ eV. The top-most curve represents Q_{inel} derived by Joshipura et al. (2006) using CSP-ic. The other theoretical data included in this figure are the single ionization results of Bartlett and Stelbovics based on the Born approximation with full orthogonalization of the continuum coulomb wave to all the occupied atomic orbitals. The experimental data-points of Freund et al. (1990) lying above the theory near the threshold indicate that there is a metastable component in the experimental atomic beams and hence the theoretical data has a Q_{ion} peak position at a higher incident energy. Joshipura et al.

(2006) report $E_p = 40$ eV in Q_{inel} and the maximum in Q_{ion} is 4.5 Å^2 around $\varepsilon_{\text{ion}} = 50$ eV. The calculated ΣQ_{exc} by Joshipura et al. shown as the lower curve in Figure 3.21 has no other data for comparison.

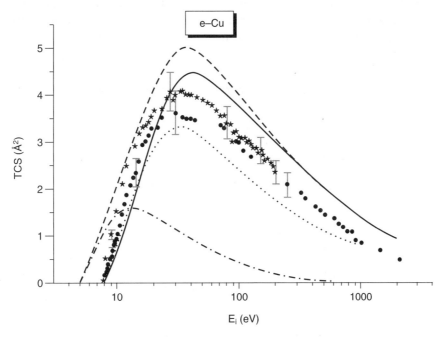

Figure 3.21 Electron scattering from copper atoms (Joshipura et al. 2006); *top-most curve:* Q_{inel}; *continuous curve:* Q_{ion}, by CSP-ic; measured data, star by Freund et al. (1990) and bullet by Bolorizadeh (1994); *the lowest curve* ΣQ_{exc} in CSP-ic

Finally, the mercury atom, on which the first ever electron impact excitation experiments were carried out by Frank and Hertz (1914) about a century ago. Earlier theoretical and experimental data on electron scattering from mercury were compiled by Zecca et al. (1996). In spite of the practical importance of such collisions, e.g., in the lighting industry, there are not many studies to report. There are a few theoretical investigations on electron impact excitation, e.g., Srivastava et al. (2009) and few electron impact ionization measurements, e.g., Almeida et al. (2001).

3.5 METASTABLE ATOMIC SPECIES

Approaching the end of this chapter, a brief detour to discuss electron scattering from excited metastable states (EMS) of atoms, is required. Here A* stands for the EMS of the atom A. Metastable species are found in many planetary and astrophysical systems as well as industrial plasmas where they play an important role in the local chemistry. Compared to the large amount of data on electron interactions with ground state (GS) atoms, studies of electron scattering from excited metastable atomic species are scarce. Box 3.2 highlights some of the general features of the atomic excited metastable states and Table 3.5 shows the first ionization energy and the radiative lifetime of a few selected metastable atoms (Pandya and Joshipura 2014, and references therein).

In this section we will first briefly review the available measured data, though rather old, on electron ionization of H and He metastable states. In each case it is interesting to compare such data with the corresponding cross sections for electron scattering from the ground sate (GS) of the same atoms.

BOX 3.2: ATOMS IN EXCITED METASTABLE STATES

In general, metastability is a well-known trait that prevails in a wide variety of physical and chemical systems. In the present context many of the atoms, molecules and ions are found to exist in excited metastable states (EMS), well understood in terms of quantum mechanical selection rules. Such states are endowed with exotic properties such as long radiative lifetimes, larger size (i.e., average radius) as well as higher polarizability and lower ionization thresholds compared to their ground-state counterparts. Since the EMS species are sufficiently long-lived, direct experimental detection is possible, in their case. Such long lifetimes also ensure that metastable species are 'chemically reactive, for example, in a collision with a target with a low ionization threshold the EMS may ionize the target, a process known as Penning ionization. He*(2^3S), with a life-time of about 7870 s (~ 2 hours), is one of the longest lived atoms in EMS. Atomic hydrogen possesses the famous 2s metastable state. For H*(2s) the ionization energy is $1/4^{th}$ of the ground state H(1s), while the average radius is about five times, that of the ground state H(1s). Static electric dipole polarizability, which is a volume effect, can be orders of magnitude larger in metastable atoms. Ionization energies and radiative lifetimes of some of the atoms in their EMS are displayed in Table 3.5. Interestingly, the EMS H* and He* have very small thresholds of ionization.

Atomic EMS can be produced by electron collisional excitation, although the relevant cross sections Q_{exc} are usually small. In natural as well as laboratory systems, metastable species may also be formed in charge exchange or in recombination processes. Although metastable species are usually present only in small quantities, they act as storehouses of energy in the medium. When this stored energy is released, it is known as quenching or deactivation of the EMS. Ionization by electron impact is one such deactivation or energy loss mechanism and hence the electron impact ionization cross sections of such metastable states are of great interest. However, it is often difficult to prepare a pure metastable target beam for electron scattering experiments; so, once again, theory is required to complete a database of electron scattering from such species.

Electron scattering from metastable species of H and He atoms has been the subject of both experimental as well as theoretical studies. Measurements on the electron impact ionization of H*(2s) were carried out, among others, by Defrance et al. (1981) in the energy range ~ 6–1000 eV, and by Dixon et al. (1975) at ~ 8–500 eV. Electron impact ionization of the 2^3S and 2^1S metastable states of helium, was studied experimentally by Dixon et al. (1976) at energies ~ 6–1000 eV. Dixon et al. (1975, 1976) and Defrance et al. (1981) also highlighted the data from previous studies.

Table 3.5 Properties of a few selected light atoms in their excited metastable states (Pandya and Joshipura 2014, and references therein)

Target species	Ionization energy (eV)	Radiative life time (s)
H*(2s)	3.40	0.14
He* (2^3S)	4.77	7870
He* (2^1S)	3.97	20×10^{-3}
N* (^{2}D)	12.15	9.4×10^4
N* (^{2}P)	10.95	12
O* (^{1}D)	11.64	110
O* (^{1}S)	9.44	0.8

In Figure 3.22 the experimentally measured Q_{ion} data on H*(2s) taken from Defrance et al. (1981) are compared with the results for electron impact ionization from the H atom in the ground state, viz., the BEB values from the NIST database and the measured data of Shah et al. (1987). The peak Q_{ion} value of the metastable target H*(2s) is at least an order of magnitude larger and the peak position occurs at a much lower energy, as one would expect.

The maximum Q_{ion} for H*(2s) can be estimated in two approximate ways, based on general physical arguments. Going back to the basic structure properties of atomic hydrogen in these two states, we have

$<r>_{1s} = 0.79$ Å, ionization threshold $I(1s) = 13.6$ eV and polarizability α_d (1s) $= 0.66$ Å^3,

and

$<r>_{2s} = 3.17$ Å, I(2s) $= 3.4$ eV and α_d(2s) $= 42$ Å^3 (Fricke 1986, Christophorou and Olthoff 2001)

The ratio of the geometrical areas of these two targets is given by,

$$\frac{(\langle r \rangle_{2s})^2}{(\langle r \rangle_{1s})^2} = 16 \text{ (approximately)}$$

Assuming the peak ionization TCS σ_{max} to be proportional to their geometrical areas, we have an estimate, called es-1,

$$\text{es-1} = \sigma_{max}(2s) = 16. \ \sigma_{max}(1s) = 10.56 \text{ Å}^2 \text{ (approximately)}$$

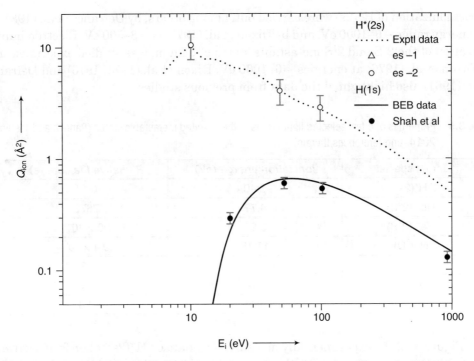

Figure 3.22 Comparisons of the ionization cross sections of the metastable species H*(2s) measured by Defrance et al. (1981); H(1s), BEB from the NIST data-base; H(1s) measured data of Shah et al. (1987); es-1 and es-2 estimates

Now, for each of these species we can consider the atomic (area-like) property defined as $P_A = \sqrt{(\alpha_d/I)}$, and assume that the peak ionization TCS is proportional to the P_A value. In the present two cases, the ratio $P_A(2s)/P_A(1s)$ is nearly 15.95, and hence we have another simple estimate, called es-2, as follows

$$\text{es-2} = \sigma_{max}(2s) = (15.95).\sigma_{max}(1s) = 10.52 \text{ Å}^2 \text{ (approximately)}$$

It is very interesting that both of these estimates are just about the same as the experimental value of Defrance et al. (1981), namely that, $\sigma_{max}(2s) = 10.5$ Å^2. Such a close agreement between two estimates and a measured value is exceptional.

In the case of helium, the measurements on electron impact ionization of the standard target He (1^{1}S) were carried out by Shah et al. (1988) while theoretical BEB Q_{ion} in this case are available from the NIST database. For He*(2^{3}S), the experimental Q_{ion} were determined by Dixon et al. (1976). Figure 3.23 compares the available results. Once again, the Q_{ion} of He*(2^{3}S) is much larger compared to Q_{ion} for the ground state He (1^{1}S). For the EMS of He the measured peak value $\sigma_{max} = 7.23$ Å^2 occurs at a much lower energy of about 12 eV.

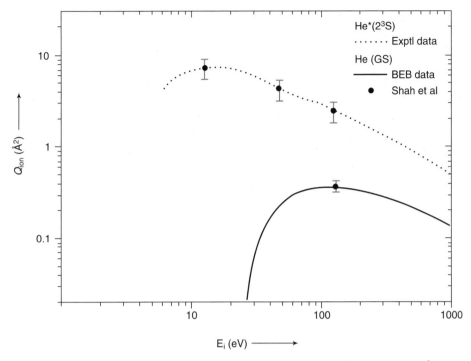

Figure 3.23 Electron impact ionization cross sections for the metastable species He*(2^3S); *dotted curve with data points*: Dixon et al. (1976); He(GS) BEB values (NIST database), He(GS); *single-point*: measured data from Shah et al. (1988)

Electron induced ionization out of metastable states is challenging for theory as can be seen from the lack of examples in the literature. Electron ionization from Ne*($1s^2 2s^2 2p^5 3s^1$) was examined using an elementary CSP-ic method by Joshipura and Antony (2001), wherein comparisons were made with an older experimental work on Ne* and also with the Ne (GS) cross sections. The results confirmed the basic observations of experiments, namely a higher Q_{ion} and correspondingly a lower peak position ε_{ion} for the metastable state.

More recently, CSP-ic theory has been extended to carry out ionization calculations for metastable species of atomic nitrogen and oxygen (Pandya and Joshipura 2014). This work was motivated by the important role played by metastable O and N atoms in the auroral phenomena observed in the Earth's polar regions. The auroral green line at 557.7 nm arises from the transition ^{1}S $\rightarrow$ ^{1}D in atomic oxygen. Formation of the metastable species takes place through collisions of energetic electrons in the upper atmosphere. Electron collisions can also ionize the existing species O*(^{1}D, ^{1}S), and this constitutes a loss mechanism for the metastable species in the environment.

Pandya and Joshipura (2014) derived and discussed electron impact ionization cross sections of the N*(^{2}P), N*(^{2}D), as well as O*(^{1}S), O*(^{1}D) and made comparisons with the respective ground states viz., N(^{4}S) and O(^{3}P). No measured Q_{ion} data are available for the O species and there appears to be only one other result on the metastable N, the BEB calculation given by Kim and Desclaux (2002).

At this stage it is appropriate to recall our discussion earlier in this chapter on electron ionization of ground state nitrogen N(^{4}S). It was mentioned with reference to Figure 3.10 that, near the ionization threshold of N(GS), the measured Q_{ion} exhibited values which were unexpectedly high. This was attributed to the presence of metastable species in the experimentally prepared atomic beam, and this was corroborated by Pandya and Joshipura (2014) as given below.

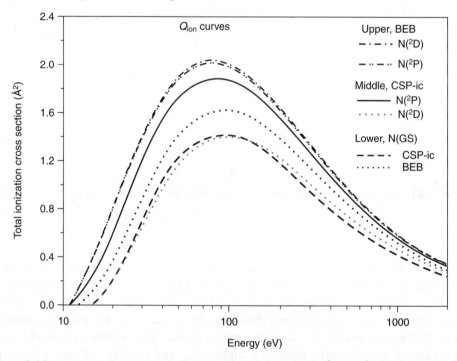

Figure 3.24 Ionization cross sections of the GS and metastable N*(^{2}P), N*(^{2}D) states of N atoms CSP-ic (Pandya and Joshipura 2014), compared with the BEB data (Kim and Desclaux 2002)

Figure 3.24 shows the CSP-ic ionization cross sections of the GS and metastable states of nitrogen atoms compared with BEB data (Kim and Desclaux 2002). The Q_{ion} of N(GS) calculated in CSP-ic, agree very well with the BEB values. Two metastable states of the N atom N*(^{2}P), N*(^{2}D) are considered here, and the Q_{ion} of both these are higher than the N(GS) values, in accordance with their excited state thresholds. The calculated Q_{ion} of these two species show appropriate differences with the peak of the upper state shifting slightly to lower energy, as one might expect. However, these features are not revealed in the BEB results of Kim and Desclaux.

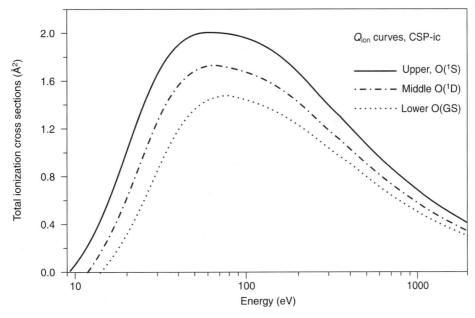

Figure 3.25 CSP-ic ionization cross sections of the GS and EMS of atomic oxygen (Pandya and Joshipura 2014)

Consider next Figure 3.25 comparing the results of CSP-ic calculations of Q_{ion} of the two oxygen metastable species O*(^{1}S) and O*(^{1}D) with that of O(GS). There appears to be no other data with which to compare these results, but the peak positions and magnitudes of the cross sections are as expected, when the threshold and ionization energies of these states are taken into account. The lowest curve in this figure shows the ground-state Q_{ion} of atomic oxygen, discussed earlier in this chapter.

Table 3.6 summarizes the CSP-ic results for Q_{ion} from the metastable and GS of atomic nitrogen and oxygen. This data provides valuable input for models of astrophysical and laboratory plasmas. Comparative numerical details on the present N and O targets, both EMS and GS, are given in Table 3.6.

Table 3.6 Impact energy at peak position, and maximum ionization cross sections of the N and O atomic species (Pandya and Joshipura 2014)

Atomic species	Peak position energy (eV)	Maximum Q_{ion} ($\mathring{A}^2$)
N(^{4}S) GS	120	1.27
N*(^{2}D)	100	1.62
N*(^{2}P)	90	1.88
O(^{3}P) GS	75	1.47
O*(^{1}D)	65	1.74
O*(^{1}S)	60	2.00

In the studies on astrophysical and laboratory plasmas, the electron impact cross sections of atoms in GS as well as EMS prove to be useful inputs. Therefore, Abdel-Nabyet al (2015) also calculated electron impact ionization cross sections for the ground and excited states of the N atom using the non-perturbative R-matrix with pseudo-states and time-dependent close coupling (TDCC) methods, as well as the perturbative distorted-wave (DW) method. The TDCC and DW results for the excited-state configurations are much larger than that for the ground configuration.

Thus, the Q_{ion} of the metastable species are larger than those of the corresponding GS. The question is; how large should they be, with respect to the GS counterpart? An approximate answer is: A general physical description can be developed by considering some of the structure properties of the target atom. Specifically, with α_d as the atomic electric dipole polarizability and I as the first ionization threshold, one finds that the quantity $\sqrt{(\alpha_d/I)}$ represents an area, which can be correlated to electron impact cross sections. In fact, the maximum ionization cross section of a target, denoted by σ_{max}, is found to be proportional to this property denoted by P_A, i.e.,

$$P_A = \sqrt{(\alpha_d/I)} \tag{3.9}$$

A connection or correlation between the static property P_A and the dynamic (i.e., energy dependent) property σ_{max} of atomic targets is quite general. Calculating the ratio of the values of 'P_A' of the EMS and the GS of an atom from their basic data and comparing this with the ratio of the maximum ionization TCS i.e., with $\sigma_{max}(EMS)/\sigma_{max}(GS)$, as obtained from CSP-ic calculations, shows that the two values are basically the same (Table 3.7). Thus knowing P_A of the EMS and the GS of an atom and electron impact ionization of the GS allows us to provide an estimated value for the Q_{ion} from the EMS of that atom.

Table 3.7 Peak cross section ratio compared with the P_A ratio for metastable oxygen

Oxygen species	$\sigma_{max}(EMS)/\sigma_{max}(GS)$	$P_A(EMS)/P_A(GS)$
O*(^{1}D)	1.18	1.11
O*(^{1}S)	1.36	1.25

3.6   GENERAL TRENDS AND CORRELATIONS IN ELECTRON–ATOM SCATTERING

In this chapter we dealt with electron scattering from a wide range of atomic targets. It is therefore pertinent to see if there is any correlation amongst the different parameters or physical properties of these atoms and their electron scattering cross sections. The first aim is to look for a connection between the maximum ionization cross σ_{max} of various ground-state atoms and their structure property viz., $P_A = \sqrt{(\alpha_d/I)}$. For the well-studied inert gas atoms He, Ne, Ar, Kr and Xe (Section 3.1), a plot of σ_{max} as a function of $\sqrt{(\alpha_d/I)}$ is depicted in Figure 3.26.

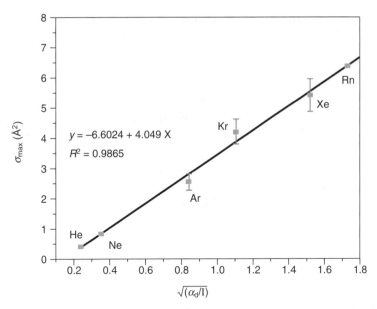

Figure 3.26   Maximum TCS σ_{max} as a function of $\sqrt{(\alpha_d/I)}$ for inert gas atoms, extrapolated to Rn atom

As the figure indicates, the maximum ionization TCS σ_{max} is linearly proportional to the quantity $\sqrt{(\alpha_d/I)}$. The regression or 'goodness' coefficient R^2 along with the linear fitting equation is given in the figure. Figure 3.26 also predicts, through extrapolation, the σ_{max} value for Rn atom, for which hardly any scattering data is available.

Another study of this kind, again for inert gas atoms, has been made by considering the polarizability α_d, which represents a characteristic volume associated with the atom. One can therefore define a 'polarizability radius' r_{pol} through $\alpha_d = 4\pi/3.(r_{pol})^3$. It differs from the radius R_{pol} of Chapter 1 by a constant factor. Accordingly, a typical area-like quantity $(r_{pol})^2$, has been defined to check for correlation with the peak cross section σ_{max} for an atomic target. That such a linear correlation exists amongst the inert gas atoms can be demonstrated, although not shown here. It appears that correlations of this kind probably exist amongst the atoms of other periodic groups, and it would be worthwhile exploring this for the halogen atoms, for which electron impact ionization has been extensively investigated.

Thus, this chapter has examined the response of a variety of atomic targets to the impact of electrons at intermediate to high energies. The CSP-ic methodology has been shown to be a good general method for estimating the total ionization and other cross sections. Observed discrepancies between calculated and measured Q_{ion} in some of the atomic targets provide an interesting discussion and in particular reveal the role of metastable states in many of the experiments. We also note that, except for inert gases, there are hardly any measurements on Q_T of other atoms. Metastable atomic targets (with the exception of H^* and He^*) are still considered to be too difficult to study in experiments and we must rely on theoretical values for many of the models and simulations in which electron scattering by metastable atoms plays an important role. Atoms in other highly excited states, the so-called Rydberg states pose even greater challenges to theory (see Dewangan 2012).

Finally, it is emphasized that most of the available electron atom scattering data are now more than twenty years old and in the twenty-first century it is still quite necessary to explore such collisions by exploiting new techniques such as atom trapping to prepare well characterized atomic targets. It must be mentioned here that electron collision cross sections measured with the use of a magneto-optical trap were reported by Shappe et al. (1995). More recently electron scattering from cold atoms along with design and operation of the AC driven magneto-optical trap was investigated by Harvey et al. (2012).

4 | ELECTRON MOLECULE SCATTERING AND IONIZATION – I

Small Molecules and Radicals

Molecules constitute a wonderful world of their own. They exhibit great diversity in terms of their chemical constituents, size, shape or structure, symmetry and other properties such as chemical isomerism. A molecule as a quantum mechanical object is full of complexities arising from its rotational, vibrational and electronic motions together with diverse ionization pathways. The multi-centered nature of molecules adds to the difficulties in developing scattering theories; however, the availability of stable molecular beams has facilitated a large number of electron–molecule experiments over a wide range of energies. In the past hundred years different electron induced processes including elastic scattering and myriad inelastic processes, such as, rotational-vibrational excitation, the formation of scattering resonances, dissociative electron attachment, neutral dissociation, electronic excitation and various ionization mechanisms have been studied extensively across a wide range of energies from zero to several keV. For a few simple molecular targets, like nitrogen, review papers (e.g., Zecca et al. 1996, Karwasz et al. 2001) appeared some years ago in literature with data recommendations being made by Itikawa et al. (1986) and updated by Itikawa (2006). A useful overview of electron–molecule interactions and their applications was given at length, in two volumes edited by Christophorou (2013).

The focus in this chapter is on processes taking place from the first ionization threshold upwards, typically in the energy range 10–2000 eV. The present chapter highlights theoretical studies on electron 'collisional' total cross sections for a wide range of common and small molecular targets, together with their radicals wherever relevant. In view of intermediate to high incident energies considered, non-spherical interactions (rotational excitations etc.) are less important. Theoretical results obtained in a semi-empirical methodology known as the Complex Scattering Potential-ionization contribution (CSP-ic) are presented here and these are compared against various available data-sets. The method of calculation begins with the concept of the electron–molecule scattering system as a Spherical Complex Optical Potential or SCOP. Doubts may be raised on this approach, as the molecule is basically non-spherical but that is justified in this case since, under normal situations, the target molecules are oriented randomly, and our interest is in total cross sections. However, non-spherical interactions, more important at lower energies, have been incorporated into calculations when practical.

A list of molecular targets discussed in this chapter, considered in a rather arbitrary sequence, is given in Box 4.1 below.

BOX 4.1

The Nitrogen molecule: N_2

The Oxygen species: O_2, O_3 and $(O_2)_2$

Oxides of Carbon: CO and CO_2

Oxides of Nitrogen: NO, NO_2 and N_2O

Halogens: F_2, Cl_2, Br_2, I_2

Lithium Hydride: LiH–a strongly polar molecule

Oxides of Hydrogen: H_2O along with OH, H_2O_2, HO_2 and $(H_2O)_2$

Hydrocarbons: CH_4 and radicals CH_x (x = 1, 2 and 3)

Hydrides of Nitrogen: NH_3 and radicals NH_x (x = 1, 2)

Sulphur related gases: SO, SO_2, H_2S

Reactive species: CN, C_2N_2, HCN and HNC; BF

Metastable species: H_2^* and N_2^*

In the discussion to follow, we have assembled molecular species in different sub-sections (in which they have common properties) starting each section, wherever possible, with a diatomic target. Critical comparisons are made with literature data and references have been made to recommended data, where these are available in a few standard cases. We begin with a very well-known and quite extensively studied molecular target N_2.

4.1　THE NITROGEN MOLECULE

Molecular nitrogen, a very important atmospheric gas, is the first choice as a target for electron scattering experiments since it is stable and inexpensive. Accordingly, the database for electron scattering from molecular nitrogen is extensive; Itikawa et al. (1986), further updated by Itikawa (2006), compiled, the then available e–N_2 cross sections and recommended data sets. Many theoretical methods have been applied to electron scattering from molecular nitrogen. Following the extensive description and success of the CSP-ic method for atoms, it has been exploited to study molecules. Joshipura et al. (2009) employed a complex potential constructed from a single-center charge density with respect to molecular mass-center and standard model forms to represent static, exchange, polarization and absorption interactions to derive the total molecular complex potential for use in the SCOP formalism. Such a representation is adequate in the intermediate and high energy regime as discussed in this book but is not sufficient for lower energy scattering formalisms. Methods of greater complexity, such as, the R-matrix and Schwinger variational methods are necessary for low energies (see Chapter 2).

The SCOP formulation yields three total cross sections Q_{el}, Q_{inel} and Q_T for a molecular target, such that $Q_T = Q_{el} + Q_{inel}$. The cross section Q_{inel} which is the sum of all admissible electronic excitations and ionizations and the cross section Q_{ion} includes single, double (and higher degrees of) ionization together with parent and dissociative ionization processes. As discussed in Chapters 2 and 3 an energy-dependent ratio function $R(E_i)$ can then be introduced,

$$R(E_i) = \frac{Q_{ion}(E_i)}{Q_{inel}(E_i)} \tag{4.1}$$

The next step is to introduce a dimensionless target-specific variable representing energy, i.e.,

$$U = E_i / I$$

with I as the first ionization energy. A continuous energy dependence of the ratio (4.1) is then expressed as

$$R(E_i) = 1 - C_1 \left[\frac{C_2}{U + a} + \frac{\ln U}{U} \right] \tag{4.2}$$

This expression, introduced on simple physical grounds, involves three parameters a, C_1 and C_2, all of which are dimensionless. They may be evaluated iteratively by invoking three conditions on $R(E_i)$ at three selected energies, as discussed in the previous chapter. In this method, a semi-empirical input used in general and for N_2 as well, is $R_p = 0.70$, at the peak inelastic position $E_p = 75$ eV for N_2. We find that the recommended data for N_2 (Itikawa 2006) yields the value $R_p = 0.73$. This small difference in R_p does not change the cross section results substantially. The three CSP-ic parameters obtained for e–N_2 scattering (Joshipura et al. 2009) are given in Table 4.1.

Table 4.1 Dimensionless parameters a, C_1, and C_2 for nitrogen molecule

a	C_1	C_2
4.93	−0.94	−6.27

In the N_2 case or in the subsequent CSP-ic calculations presented in this chapter no attempt is made to adjust R_p or any other parameter to force an agreement with experimental or other data, and as will be seen, the CSP-ic method achieves a reasonable degree of success for a large number of atomic/molecular targets.

It must be noted that, in general, the cross sections known most confidently for a target are the total (complete) cross sections Q_T and the total ionization TCS Q_{ion}. It should also be noted that two of the theoretical quantities derived by CSP-ic method viz., the inelastic TCS Q_{inel} and electronic-sum ΣQ_{exc} are not available as such in any other publications on N_2 or, for that matter, for any other target. However in the case of nitrogen, data-sets on Q_{inel} and ΣQ_{exc} can be deduced by employing the recommended data values of Itikawa (2006) in the definition,

$$Q_{inel} = Q_T - Q_{el} \tag{4.3}$$

and,

$$\Sigma Q_{exc} = Q_{inel} - Q_{ion} \tag{4.4}$$

Thus for the N_2 molecule, using equation (4.3) together with the recommended data (Itikawa 2006) at the peak energy E_p yields the value $Q_{inel} = 3.37$ Å^2, quite close to the presently calculated value of 3.36 Å^2. Similarly using equation (4.4), ΣQ_{exc} at E_p deduced from the recommended data is 0.90 Å^2 while the CSP-ic calculated value is 0.75 Å^2. However the calculations of Joshipura et al. did not include processes like quadrupole or rotational excitation, neutral dissociation, etc.

Figure 4.1 presents graphical plots of the calculated TCS Q_T, Q_{el}, Q_{inel}, Q_{ion} and ΣQ_{exc} showing several different sets of data for e–N_2 scattering. Reviewing each of the cross sections in turn for N_2, there are six sets of measured data on Q_T, and these are from Dalba et al. (1980), Blaauw et al. (1980), Hoffman et al. (1982), Garcia et al. (1988), Nogueira et al. (1985) and Karwasz et al. (1993). There is very good agreement between the calculated CSP-ic values of Joshipura et al. (2009) and the six sets of experimental data while the theoretical values of Jain and Baluja (1992) are seen to be on higher side. The experimental measurements correspond to different ranges of incident energy. However, a notable feature for N_2 is a broad bumpy structure in the theoretical Q_T around 50-60 eV, near the peak position $E_p = 75$ eV of inelastic TCS Q_{inel}, where it should be noted that $Q_{el} > Q_{inel}$. Note here the significant difference between the calculated Q_{el} and compared data below 100 eV. For elastic cross sections Q_{el} the CSP-ic is compared with measured data given by Herrmann et al. (1976), Dubios and Rudd (1976), Shyn and Carrignan (1980) and Trajmar et al. (1983), and also with the recommended data of Itikawa (2006). The calculated Q_{el} and Q_T merge at the lower end of energy, as expected. The Q_{ion} determined in CSP-ic exhibits a broad peak near 100 eV and from Itikawa (2006) 74% of the cumulative cross section Q_{ion} arises from the parent ion N_2. At high energies the elastic cross section Q_{el} and Q_{ion} are approximately the same and dominate the TCS Q_T.

For clarity the TCS corresponding to cumulative ionization as well as electronic excitations are shown separately in Figure 4.2. Here, the results of Hwang et al. (1993) are from the BEB theory while the experimental data sets are from Rapp et al. (1965), Krishnakumar and Srivastava (1990) and Straub et al. (1996). In the BEB method, the ionization cross section per molecular orbital (denoted by σ_{BEB}) is expressed in terms of basic molecular orbital parameters, and the total ionization cross section is obtained as the sum over the orbitals. The agreement of CSP-ic Q_{ion} calculations with these data is seen to be excellent. At about 30 eV (close to 2I, with I = 15.58 eV for N_2), the ionization cross section equals the electronic-sum ΣQ_{exc}. Ionization corresponds to transitions to continuum scattering channels, which dominate progressively over the discrete excitations, typically above 30 eV.

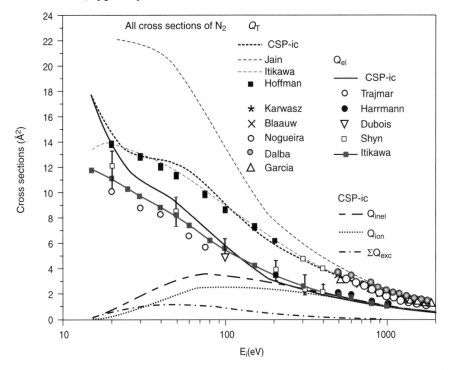

Figure 4.1 Total cross sections of e–N_2 collisions, see text for references (Joshipura et al. 2009)

The cumulative total cross section for neutral dissociation of N_2, is denoted by Q_{NDiss}, corresponding to the electron impact process

$$e + N_2 \rightarrow N + N + e \tag{4.5}$$

In this process, the threshold energy is the bond-dissociation energy (about 9.8 eV) and both the outgoing N atoms may be assumed to be in the ground state. Unlike the H_2 molecule (discussed in Chapter 2), there is no directly repulsive state of N_2, from which de-excitation will result in neutral dissociation. The potential energy diagram for N_2 is much more complicated than for H_2, and there are only a few repulsive states available. In view of a large number of electronically excited states of N_2, we take advantage of a cumulative quantity viz., the electronic-sum $\Sigma Q_{exc}(i \rightarrow f)$ introduced in the CSP-ic formalism. This quantity, deduced from Q_{inel}, represents excitations to all accessible final electronic states. Since we are interested in neutral dissociation, we must subtract from this quantity, the total emission cross sections Q_{emiss} corresponding to the major emission lines of N_2, as discussed in Pandya and Joshipura (2010, 2014). Thus, ignoring rather small cross sections for excitations to electronically metastable states of N_2, we have

$$Q_{NDiss} = \Sigma Q_{exc} - Q_{emiss}$$

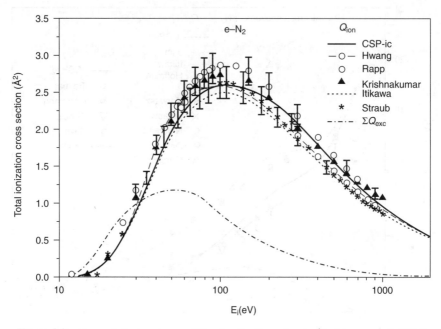

Figure 4.2 Q_{ion} and electronic-sum ΣQ_{exc} for the N_2 molecule (Joshipura et al. 2009)

To determine the effective neutral dissociation cross section, we must also consider the electron induced dissociative *ionization* of N_2, which produces one N atom per event via the following process

$$e_i + N_2 \rightarrow N + N^+ + e_{sc} + e_j \tag{4.6}$$

where e_i and e_{sc} indicate the incident and the scattered electron respectively, while e_j stands for the electron ejected from the target. Let us denote the total cross section for dissociative ionization (4.6) by Q_{dion}. Lindsay and Mangan (2003) had determined recommended values of partial and total ionization cross sections for N_2 (see also Itikawa 2006). The dissociative ionization cross section Q_{dion} used by Pandya and Joshipura is that of Lindsay and Mangan (2003). By appropriately including the contribution of Q_{dion} the effective total neutral dissociation cross section for N_2 was calculated, and is shown by the continuous black curve in Figure 4.3.

Winters (1966) determined the total dissociation cross section for neutral products in N_2, and later Cosby (1993) obtained this by directly detecting the fragment pair N + N. Cosby (1993) also recommended data for neutral dissociation of N_2. Accordingly two sets of data by Cosby were included by Pandya and Joshipura, from where Figure 4.3

has been reproduced. Furthermore, the values denoted by 'corrected ΣQ_{exc}' in this figure correspond to the difference ($\Sigma Q_{exc} - Q_{emiss}$). There is a considerable spread in the measured datasets, and therefore the present estimate of effective Q_{NDiss} (continuous black curve in Figure 4.3) can be considered to be satisfactory.

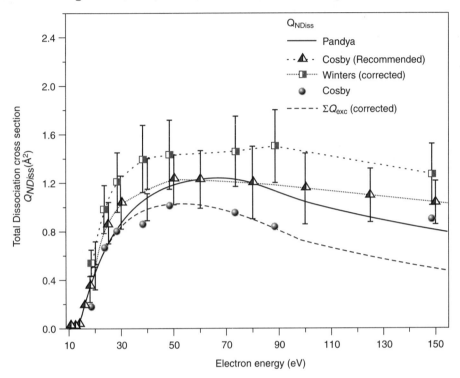

Figure 4.3 Cumulative total neutral dissociation cross section of N_2; continuous curve-effective Q_{NDiss} and 'corrected' ΣQ_{exc} (Pandya and Joshipura 2010, 2014); *squares* – Winters (1966); *bullets and triangles* – original and recommended data Cosby (1993)

Electron scattering with the nitrogen molecule initially in the electronically excited metastable state is considered later on in this chapter.

4.2 OTHER DIATOMIC AND WELL-KNOWN TARGETS

In this section we will highlight the application of the CSP-ic methodology to other diatomic and well-known molecular targets. For the target sequences considered below, the first member is a diatomic target if appropriate.

4.2.1 Oxygen species $O-O_2-O_3-(O_2)_2$

The element oxygen is unique in the sense that it exists in monoatomic, diatomic as well as triatomic forms, while a dimer $(O_2)_2$ is formed under pressure in specific environments. General aspects on dimers are highlighted in Box 4.2 later in this chapter. It would be desirable to make a comparative study of the various total cross sections of the oxygen species $O-O_2-O_3-O_4$, determined within a common general formalism. For the standard O_2 target, apart from the old review by Zecca et al. (1996), recommended cross sections were given by Itikawa (2009). An R–matrix formalism of low energy electron scattering with O_2 was studied by Singh and Baluja (2014), who also reported the BEB model ionization cross sections, available in the R–matrix package.

Theoretical CSP-ic investigations on the $O-O_2-O_3-O_4$ sequence were reported more than a decade ago in Joshipura et al. (2002). For calculations on O_2 and O_3 molecules, a single center charge density was developed with the respective center of mass as the origin, and hence an energy-dependent 4-component spherical complex potential was constructed. Theoretical studies on the structure of loosely bound Van der Waals dimers such as $(O_2)_2$ were carried out by Bussery-Honvault and Veyret (1998). The O_4 molecule, regarded as the dimer of O_2 possessing a parallel coplanar D_{2h} geometry, eventually dissociates into $O_2(^1\Delta_g) + O_2(^1\Delta_g)$. The oxygen dimer exhibits a rather large inter-molecular separation of about 3.17 Å, as against the 'monomer' bond length of $R_{O-O} = 1.21$ Å; therefore, there is a rather small overlap of the electron charge-clouds of the two constituent oxygen molecules in $(O_2)_2$. Hence each O_2 can be considered as an approximately independent scatterer, such that a simple Additivity Rule can be applied. In a simple high energy Additivity Rule (AR) the total cross section for a molecule XY is given approximately by

$$Q_{add}(XY) = Q(X) + Q(Y) \tag{4.7}$$

where, $Q(X)$ and $Q(Y)$ are the corresponding TCS of constituents (atoms or functional groups) X and Y treated as 'free' or isolated. The AR in general leads to an overestimation of the cross sections, unless the impact energies are high enough. The AR is expected to hold better for a dimer, and hence Joshipura et al. (2002) adopted a simple expression, $Q_{ion}(O_4) = 2^\beta \cdot Q_{ion}(O_2)$, with $\beta = 0.8$ to account for the bonding in the dimer.

CSP-ic results for Q_T and Q_{ion} in the oxygen sequence $O-O_2-O_3-O_4$ together with the available comparisons are shown graphically in Figures 4.4 and 4.5. Figure 4.4 depicts, from bottom to top, the Q_T of the four targets O, O_2, O_3 and O_4. Comparisons with two experimental sets of data from Dababneh et al. (1988) and Dalba et al. (1980) have been included for O_2, for which good agreement is seen. The TCS increases with the number of electrons along the target sequence. In the case of the ozone molecule, Figure 4.4 shows a good agreement between the CSP-ic results and the high energy (< 300 eV) measurements of Q_T by Pablos et al. (2002).

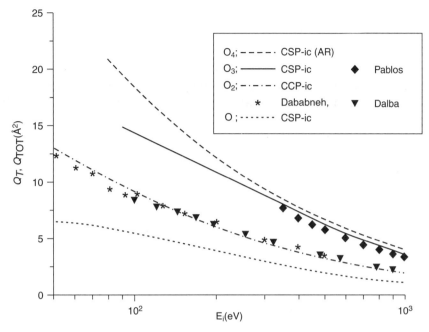

Figure 4.4 Q_T for the target sequence O, O_2, O_3, and O_4 from bottom to top; for O_2 the measured data are from Dababneh et al. (1988) and Dalba et al. (1980); for O_3 the measurements are from Pablos et al. (2002)

Figure 4.5 displays the CSP-ic and other ionization cross sections for this sequence of four targets. The theoretical CSP-ic data and experimental data of Thomson et al. (1995) for atomic oxygen are in good agreement. For O_2 the BEB theory results are from Kim et al. (1997, see also NIST database) while the measured data are from Krishnakumar and Srivastava (1992); both these agree well with the CSP-ic data. The contribution from ionization and excitation channels in O_2 is seen to be equal at about $E_i = 2I \approx 25$ eV ($I = 12.07$ eV). In Table 4.2, the CSP-ic values are compared with the recommended and other data at 100 eV. CSP-ic Q_T data is in good agreement with the recommended values of Itikawa (2009) while the recommended data for Q_{ion} agrees well with both BEB and CSP-ic values. However the Q_{el} derived in CSP-ic is notably higher than the recommended data, which is based on experimental results.

Due to the difficulties in preparing molecular beams of ozone there have been few experimental studies of O_3; one of the earliest theoretical calculations is that of Joshipura et al. (2002) whose values for Q_T are in good agreement with the experimental data of Pablos et al. (2002); Figure 4.6. The only measured Q_{ion} data for ozone are those of Newson et al. (1995) and these are much lower than the calculated CSP-ic values and the BEB data (Kim et al. 1997, NIST database). An interesting outcome of the ionization calculations on O_3 and O_2 is that, at 100 eV, we have

$$\frac{Q_{\text{ion}}(O_3)}{Q_{\text{ion}}(O_2)} = 1.52 \tag{4.8}$$

Probably the first results on electron collisions with the dimer $(O_2)_2$ were reported by Joshipura et al. (2002). Preliminary estimates on the electron-induced Q_{ion} of this dimer derived in the independent 'scatterer' approximation (or the AR) are shown in the top curve of Figure 4.5. All the total ionization cross sections given in this figure are consistent with the respective target properties of the oxygen species, namely ionization threshold, target-size (molecular bond lengths) and the total number of target electrons.

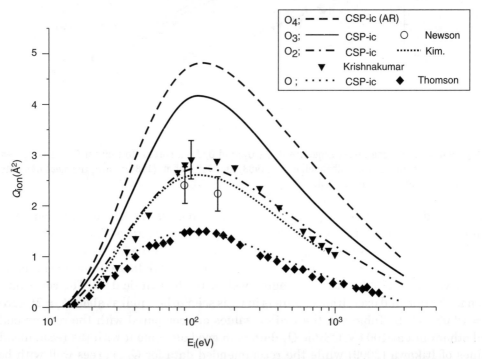

Figure 4.5   Ionization cross sections of $0-0_2-0_3-0_4$ sequence from bottom to top (Joshipura et al. 2002); atomic 0, *dot*: CSP-ic, experimental data Thompson et al. (1995); *middle curves* 0_2, *dash-dot*: CSP-ic; *dot*: Kim et al. (1997), *inverted triangles*: measured data – Krishnakumar and Srivastava (1992); for 0_3 *continuous curve*: CSP-ic, *open circle*: measurements – Newson et al. (1995); *uppermost curve*: 0_4

Table 4.2   100 eV cross sections (in $Å^2$) for 0_2 molecule; (a) recommended data: Itikawa (2009), (b) Joshipura et al. (2002), (c) BEB (NIST database)

Q_T		Q_{el}		Q_{ion}		
(a)	*(b)*	*(a)*	*(b)*	*(a)*	*(b)*	*(c)*
8.68	8.90	4.78	5.30	2.43	2.75	2.62

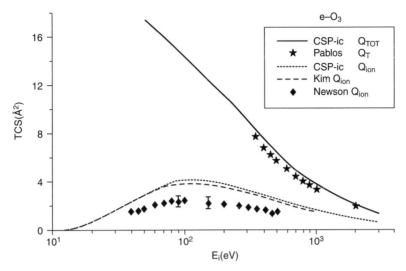

Figure 4.6   Total (complete) and ionization cross sections for ozone molecule; Q_{TOT} *uppermost curve* CSP-ic; *stars*: Pablos et al. (2002): Q_{ion}, *dots*: CSP-ic, *dashes*: BEB (Kim et al. 1997, see also NIST database), *diamonds:* Newson et al. (1995)

Figures 4.7a and 4.7b are in the form of bar-charts, and they present the theoretical values of different total cross sections from oxygen and ozone molecules at the peak of the Q_{ion} which is 100 eV. The relative contributions for each process are remarkably similar in the two targets.

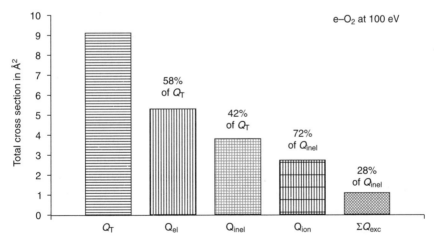

Figure 4.7(a)   Bar chart showing the relative contribution of various electron collision processes in O_2 at the maximum of Q_{ion}, 100 eV (Joshipura et al. 2002)

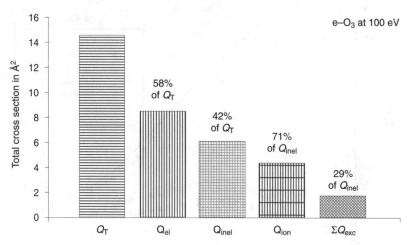

Figure 4.7(b)　Bar chart showing the relative contribution of various electron collision processes in O_3 at the maximum of Q_{ion}, 100 eV (Joshipura et al. 2002)

4.2.2　Carbon monoxide (CO) and carbon dioxide (CO_2)

The carbon monoxide molecule is second only to H_2 in abundance in interstellar space and is a major species in the atmospheres of Venus and Mars, while it is infamous as a pollutant in the terrestrial atmosphere. Recommended datasets for electron scattering from CO have been reported recently by Itikawa (2015).

The bond length in the CO molecule is R_{C-O} = 1.13 Å, while its first ionization threshold is 14.01 eV and polarizability α_0 is 1.95 Å^3. Q_T, Q_{el}, Q_{inel}, Q_{ion} and ΣQ_{exc} for e–CO collisions are shown in Figure 4.8. CO is one of those targets for which the CSP-ic theory (Pandya 2013 and Vinodkumar et al. 2010) agrees very well with other data for almost all the TCS over the entire energy range. While no experimental data is available for the electronic-sum ΣQ_{exc}, it is notably lower than Q_{ion}.

The total (complete) Q_T CSP-ic cross section (Pandya 2013) agrees well with experimental data of Kanik et al. (1992), Karwasz et al. (1993), Garcia et al. (1990) and Kwan et al. (1983), and shows a broad structure near 50–60 eV (Figure 4.8). Such a structure is also seen in Q_T of the isoelectronic molecule N_2 (Figure 4.1). CSP-ic Q_{el} of CO is tested against the rather old experimental results of DuBois and Rudd (1976) and Bromberg (1970) and, in general there is good agreement.

So how does the theory compare with the recommended data on CO published recently by Itikawa (2015)? At 100 eV, Q_T in CSP-ic is 9.23 Å^2, in excellent agreement with the recommended value 9.27 Å^2. However, the CSP-ic Q_{el} is 5.77Å^2 while the recommended integral elastic cross section is 4.2 Å^2. The elastic cross section is equal to the inelastic cross section at about 150 eV (Figure 4.8). The CSP-ic ionization results (Figure 4.8) match those derived using the BEB method (Hwang et al. 1996, see also NIST Database).

Fairly good agreement is found between CSP-ic Q_{ion} and the earliest comprehensive ionization experiments of Rapp and Englander-Golden (1965) and the Q_{ion} of Orient and Srivastava (1987). At the maximum of Q_{ion} viz. 100 eV the CSP-ic value of Q_{ion} i.e., 2.60 Å^2 is in accord with Itikawa's recommended value of 2.64 Å^2. The lowest curve in Figure 4.8 exhibits the electronic – sum for ΣQ_{exc} in e–CO collisions, which is important in view of the lack of any relevant experimental data.

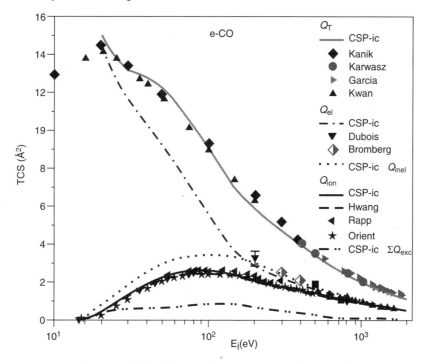

Figure 4.8 TCS plot for e–CO scattering (Pandya 2013)

Now let us review the electron scattering from carbon dioxide molecule. The most recent data compilation and recommendation for various electron collision processes with carbon dioxide was published over a decade ago by Itikawa (2002). The recommended data for CO_2 as given by Itikawa in Å^2 are $Q_T = 12.6$, $Q_{el} = 7.55$ and $Q_{ion} = 3.64$. Total (complete) cross sections calculated in CSP-ic are not given here, but ionizing collisions are considered for this target. As shown in Figure 4.9, the CSP-ic calculations of Q_{ion} with an input $R_p = 0.70$ (Pandya 2013) near the peak region are slightly lower than those derived from the BEB method (Hwang et al. 1997, see also NIST Database) and the measurements of Straub et al. (1996). The theoretical results of Vinodkumar et al. (2010) are very similar to the those of Pandya (2013).

Of relevance here, is a brief mention of electron scattering from the dicarbon C_2, an exotic molecule having a relatively larger bond length of 1.24 Å, along with first ionization threshold $I = 11.87$ eV and $\alpha_0 = 3.0$Å^3. It is found to exist in carbon vapour, for example, in electric arcs, in comets, stellar atmospheres, the interstellar medium and in blue hydrocarbon flames. A recent calculation of e–C_2 scattering using CSP-ic method has been reported by Naghma and Antony (2013), who found the maximum Q_{ion} of C_2 to be 3.67 Å^2 at 80 eV.

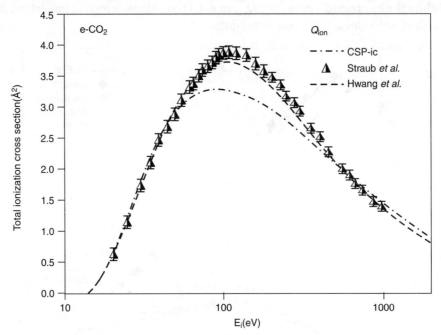

Figure 4.9 Q_{ion} in e–CO_2 scattering, compared with BEB calculations (Hwang et al. 1997) and measured data of Straub et al. (1996)

4.2.3   Nitric oxide (NO), nitrogen dioxide (NO$_2$) and nitrous Oxide (N$_2$O)

Oxides of nitrogen are important minor constituents in the terrestrial atmosphere. Nitrogen oxides NO_x ($x = 1, 2$) are produced during lightning and electrical discharges in the atmosphere. N_2O is emitted from both biological sources and from human activities and is a powerful greenhouse gas. A review of earlier work on these molecules may be found in the old data compilations of Zecca et al. (1996) and Karwasz et al. (2001).

Consider first the nitric oxide (NO) molecule. Figure 4.10, reproduced from Joshipura et al. (2007), shows data for Q_T, Q_{el}, Q_{inel}, Q_{ion} and ΣQ_{exc}. In the energy range 10–2000 eV the small dipole moment of the NO molecule may be ignored and the CSP-ic method can be used to derive the set of cross sections (Joshipura et al.).

For the nitric oxide molecule, Dalba et al. (1980) measured Q_T in the range 100–1600 eV while lower energy measurements were reported by Szmytkowski and Maciag (1991). The CSP-ic results for Q_T agree very well with the data of Dalba et al. above 100 eV but the theoretical results are increasingly higher than the measurements of Szmytkowski and Maciag at energies below 90 eV (Figure 4.10). In order to investigate this discrepancy a comparable dataset was prepared by adding the elastic cross sections Q_{el} of Fujimoto and Lee (2000) to the experimental ionization cross sections Q_{ion} of Iga et al. (1996). Joshipura et al. found that $Q_{el} + Q_{ion}$ was somewhat higher than the measured Q_T values of Szmytkowski and Maciag and would be still higher with the inclusion of electronic excitation. The overestimation of CSP-ic at lower energies may be attributed to limitations in the model exchange and polarization potentials adopted in the calculations. On the other hand, although NO is a weakly polar molecule, its dipolar nature can introduce experimental uncertainties in the measurements of Q_T especially at lower energies. The integrated Q_{el} of Kubo et al. (1981), shown by an open triangle in Figure 4.10 at 20 eV, is seen to lie close to the experimental Q_T of Szmytkowski and Maciag. Theoretical elastic cross sections Q_{el} plotted in the middle of Figure 4.10 show the expected behaviour, in the sense that they tend to lie below the ionization cross sections beyond the maximum of Q_{ion}.

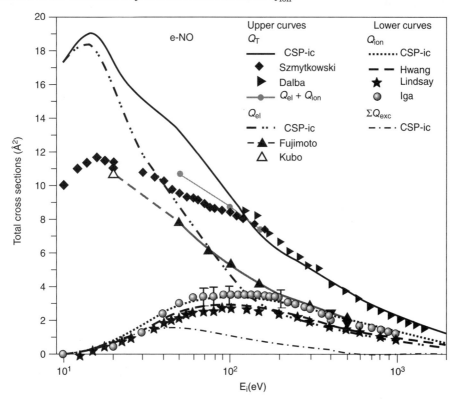

Figure 4.10 Electron collision cross sections with the NO molecule adopted from Joshipura et al. (2007)

CSP-ic values of Q_{ion} are in a very good agreement with the measurements made by Iga et al. (1996). However, the BEB Q_{ion} of Hwang *et al.* (1996) are closer to the measurements of Lindsay et al. (2000). One can ask, which of these two data sets is correct?

The open - shell $^2\pi$ NO molecule has a relatively low ionization threshold at 9.26 eV, so Q_{ion} are expected to be high in the peak region as indicated by the CSP-ic results.

The lowest curve in Figure 4.10 depicts the summed total electronic excitation cross sections ΣQ_{exc} for NO resulting from CSP-ic. Brunger (2007) reviewed this data using their own dataset for electron impact excitation and concluded that the excitation-sum ΣQ_{exc} data of Joshipura et al. is on the higher side. Joshipura et al. (2007) in their reply to Brunger et al. accepted their comments, considered the sum-check and suggested that there was little agreement between theory and experiment! The inelastic or absorption cross sections Q_{inel} for NO were also calculated by Fujimoto and Lee (2000). However, their values of Q_{inel} are lower than the Q_{ion} data, and this cannot be true.

Figure 4.11 shows the relative contributions of the TCS for NO at the maximum of Q_{ion} once again at 100 eV. At this energy Q_{el} is almost equal to Q_{inel} while Q_{ion} is 76% of the Q_{inel}.

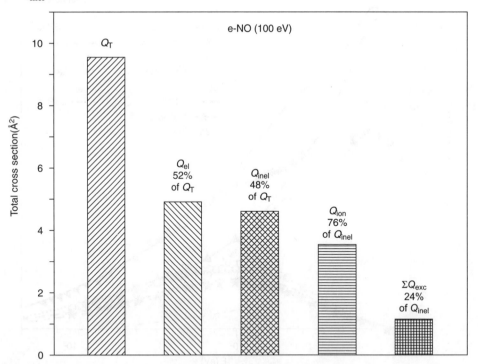

Figure 4.11   Relative contribution of various TCS in e–NO collisions at 100 eV (Joshipura et al. 2007)

A recent review-cum-data paper on e–NO collisions has been published by Itikawa (2016), in which the data recommended for Q_T and Q_{el} are drawn from earlier experiments. As can be seen from Table 4.3, CSP-ic results do not agree very well with the recommended data (Itikawa 2016).

Table 4.3   CSP-ic TCS (in Å²) of e–NO scattering, at 100 eV, with the corresponding data from Itikawa (2016) given in brackets

Q_T	Q_{el}	Q_{ion}	ΣQ_{exc}
9.46 (8.22)	4.90 (5.36)	3.46 (2.75)	1.10

Joshipura et al. (2007) also studied electron scattering from the other nitrogen oxides N_2O, NO_2, NO_3 and N_2O_5. However, for the purpose of brevity, we have summarized in a single plot (Figure 4.12) the CSP-ic ionization cross sections of NO, N_2O and NO_2 targets. The relative values of peak Q_{ion} in this plot are as expected from their molecular properties. The theoretical results for NO_2 have been included in the data paper on Gaseous Electronics, by Govinda Raju (2011).

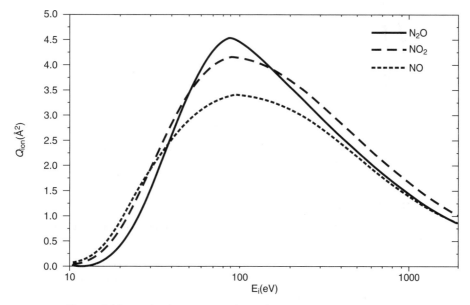

Figure 4.12   Ionization cross sections of NO, N_2O, and NO_2 molecules, adapted from Joshipura et al. (2007)

Before moving to heavier molecules with larger numbers of target electrons it is interesting to compare theoretical Q_{ion} for a few of the light atoms and molecules. This comparison is illustrated in Figure 4.13, reproduced from the application-oriented paper of Haider et al. (2012). The general shape of the cross sections and the magnitudes are very similar for isoelectronic CO and N_2. Those of O and N are also very similar but interestingly those of NO and CO_2 are also seen to be similar even though CO_2 has more electrons than NO.

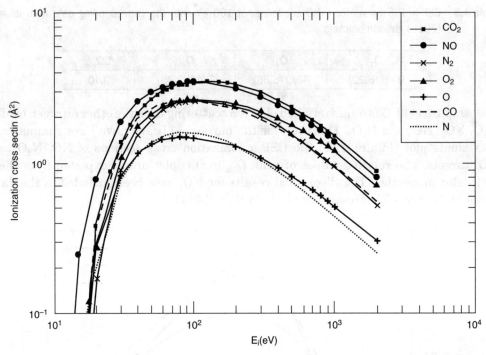

Figure 4.13   Comparison of the calculated Q_{ion} of well-known light atoms and molecules from Haider et al. (2012)

4.2.4   Halogens: F_2, Cl_2, Br_2, and I_2

The halogens are well known for their reactivity, fluorine being the most reactive. Ionizing collisions of electrons with the sequence F_2, Cl_2, Br_2, I_2 were examined theoretically by Joshipura and Limbachiya (2002). Figure 4.14 shows Q_{ion} for e–F_2 ionization. The data of Joshipura and Limbachiya are much higher than the measurements of Rao and Srivastava (1996) and the earlier data of Center and Mandel (1972) and Stevie and Vasile (1981). Also shown in Figure 4.14 are the results from the Deutsch–Maerk (DM) method (Deutsch et al. 2000) and the semi-empirical Defect Concept (DC) method of Probst et al. (2001); both of these results are again higher than the experimental data. This difference between theory and experiment is observed for many other molecules containing fluorine atom(s) as discussed later in this book.

Figure 4.15 summarizes the status of electron impact ionization for the Cl_2 molecule. The experimental data are from Stevie and Vasile (1981), while the other theoretical data are from Deutsch et al. (2000) and Probst et al. (2001). CSP-ic values of Joshipura and Limbachiya (2002) are in good agreement with the recommended data of Christophorou and Olthoff (1999). The DM values of Deutsch et al. (2000) rise more slowly and are lower than the experimental and CSP-ic at higher energies.

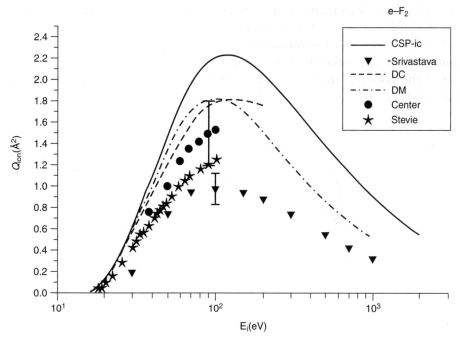

Figure 4.14   Electron impact ionization cross section of F_2 (Joshipura and Limbachiya 2002)

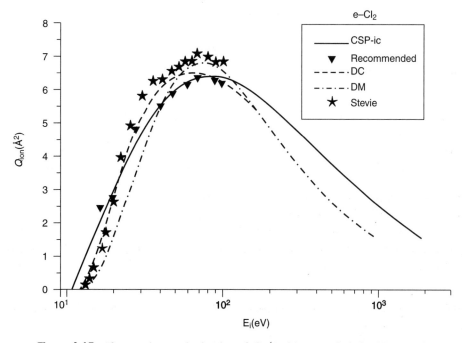

Figure 4.15   Electron impact ionization of Cl_2 (Joshipura and Limbachiya 2002)

Figure 4.16 shows three theoretical calculations of electron impact ionization of Br_2, namely the CSP-ic results of Joshipura and Limbachiya (2002), the DM method of Deutsch et al. (2000) and the semi-empirical DC method of Probst et al. (2001). The maximum magnitude of the cross section is in agreement for all three methods but the position of the peak and the shape of the cross section are seen to be different for the three methods.

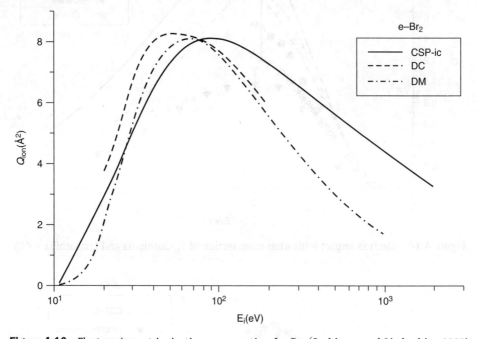

Figure 4.16 Electron impact ionization cross section for Br_2 (Joshipura and Limbachiya 2002)

Finally, in Figure 4.17 the CSP-ic results (Joshipura and Limbachiya 2002) for electron impact ionization of the I_2 molecule are compared with those of Deutsch et al. (2000) and Probst et al. (2001). Once again the shapes of the cross sections are different but the CSP-ic and DM methods agree that the peak energy of the cross section is at 50 eV.

More recently, Ali and Kim (2008) reported BEB calculations on two of the halogen molecules Br_2 and I_2. Their results are in good agreement with the CSP-ic results up to the ionization peak but thereafter the BEB cross section falls more rapidly. As a sample comparison of these two theoretical results Table 4.4 compares Q_{ion} for bromine and iodine molecules at 60 eV.

Table 4.4 Q_{ion} (in $Å^2$) for the two diatomic halogens, in CSP-ic and in BEB (Ali and Kim 2008) at 60 eV energy

Br_2		I_2	
CSP-ic	*BEB*	*CSP-ic*	*BEB*
7.75	9.01	12.35	12.06

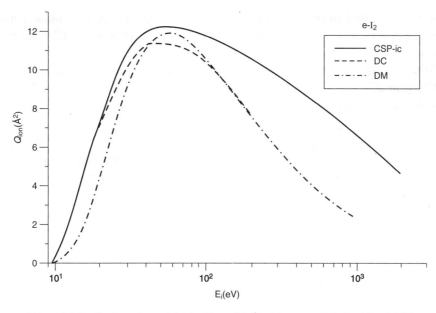

Figure 4.17 Electron impact ionization of I_2 (Joshipura and Limbachiya 2002)

4.2.5   Lithium Hydride, LiH – a strongly polar molecule

Electron scattering with the light diatomic molecule LiH has not received much attention. LiH and LiH$^+$ were the major molecular species in the early universe and they played a key role in its early chemistry while interacting with the cosmic background radiation (Gianturco and Giogi 1996, Lepp et al. 2002, Bodo et al. 2003). There is special interest in this target since it is a strongly polar molecule (D = 2.31 au) with a relatively large bond length of 1.595 Å and a high polarizability of about 4 Å^3, together with a rather low ionization threshold I = 7.9 eV. All these properties suggest high cross sections of electron scattering. Classically speaking, a molecular dipole interacting with the electric field of an incident electron experiences a torque which causes rotation. Polar molecules respond strongly to slow electrons and exhibit very large rotational excitation cross sections Q_{rot}. In this case, the Born-dipole calculations, though easy to carry out, would overestimate the cross section in view of the large value of D. Therefore in theoretical investigations on LiH, Shelat et al. (2011) used a realistic dipole potential (see Chapter 2, Section 2.2) to evaluate the total rotational excitation cross sections Q_{rot} corresponding to transition from $J = 0$ to $J'=1$. At intermediate electron energies the Q_{rot} of polar targets provides a significant contribution to the total overall scattering cross section and hence in the framework of CSP-ic theory the non-spherical contribution is included by defining the grand total cross section Q_{TOT} viz.,

$$Q_{TOT} = Q_T + Q_{rot}$$

$$= (Q_{el} + Q_{rot}) + Q_{inel} \tag{4.9}$$

Thus, Q_{rot} is added to the Q_{el} calculated in CSP-ic to obtain rotationally inclusive total elastic cross sections. In Figure 4.18 all the major TCS of e–LiH scattering are plotted. Here, the symbols Q_{rot} (1) and Q_{TOT} (1) stand for the results in the realistic 'cut-off' dipole model, where the cut-off distance R_d in the model potential (see equation 2.6 of Chapter 2) is chosen to be the half of the bond length in LiH molecule. Q_{TOT} evaluated with a point-dipole potential tend to be quite large at low energies.

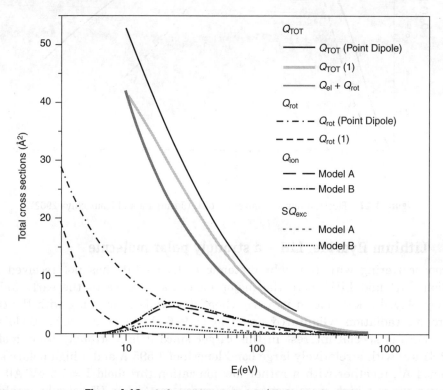

Figure 4.18 Q_T for *e*–LiH scattering (Shelat et al. 2011)

A comparison of the two dipole models is shown by the two curves of rotational excitation TCS plotted separately in the left lower part of Figure 4.18. Low energy R-matrix calculations of e–LiH scattering were also carried out by Antony et al. (2004) with and without the Born corrections and are similar in magnitude and shape to the point dipole approximation.

For lithium hydride, while the grand total and the elastic cross sections are quite large at low to intermediate energies, the ionization cross section is relatively small, Figure 4.18. In view of the lack of experimental data, two models of theory were considered (Shelat et al. 2011). In Figure 4.18, model A shows Q_{ion} with CSP-ic input $R_p = 0.7$ while model B corresponds to the input $R_p = 0.75$.

4.3 THE WATER MOLECULE, ITS DERIVATIVES OH, HO$_2$, H$_2$O$_2$, AND THE WATER DIMER (H$_2$O)$_2$

4.3.1 H$_2$O, OH

Water is the simplest of the compounds formed by oxygen and hydrogen and the third most abundant molecule in the universe next only to H$_2$ and CO. Water is one of the most ubiquitous molecules playing an important role in atmospheric chemistry, radiation induced damage within cellular systems and in various plasmas. It is often found in its dimer form (for example in biological systems) or is dissociated (e.g., by photolysis in the terrestrial atmosphere or by electron impact in atmospheric-pressure plasma) to form OH radicals. A close chemical relative of OH is the hydroperoxyl radical, HO$_2$ which is a major source of tropospheric ozone (Martinez et al. 2003). OH and HO$_2$ together with H$_2$O$_2$ initiate and participate in almost all of the complex chemical pathways of the atmosphere particularly in the troposphere (Logan et al. 1981, Ehhalt et al. 1991, Monks 2005).

Table 4.5 shows at a glance the structural properties of the five target species made up of O and H atomic constituents. Since all the five systems possess a permanent dipole moment an appreciable contribution is expected from Q_{rot} at lower impact energies. All these molecules exhibit an almost equal O – H bond length. These properties help to gain insight into the response of these molecular targets to the impinging electrons.

Table 4.5 Structural properties of the present sequence of targets (from the CCCBDB database)

Molecule	First ionization energy eV	O – H bond length Å	O – O bond length Å	Dipole moment au
H$_2$O	12.61	0.96	-	0.72
OH	13.02	0.97	-	0.60
HO$_2$	11.35 Vertical IE 11.54	0.97	1.331	0.78
H$_2$O$_2$	10.58 Vertical IE 11.70	0.95	1.475	0.69
(H$_2$O)$_2$	11.71	0.96	2.98	0.98

BOX 4.2: DIMERS AND CLUSTERS – EXOTIC FORMS OF MATTER

Inert gases are monoatomic gases, i.e., they exist normally as 'monomers'. On the other hand, the well-known homonuclear diatomic molecules such H_2, O_2, N_2, etc., are formed essentially because the molecular formation from respective identical atoms leads to decrease of total energy and this provides stability in the bound system. Molecules such as H_2, O_2, and N_2, etc., are formed by covalent bonding, a concept developed by G. N. Lewis almost a hundred years ago. The bond dissociation energy D_0 of such molecules is relatively high. The dissociation energies of H_2, O_2, and N_2 are 4.5, 5.2 and 9.8 eV respectively, and their bond lengths are 0.74, 1.21 and 1.10 in Å respectively. These molecules are stable at normal temperatures.

Dimers and clusters are relatively loosely bound systems. A dimer is formed by two structural units, called monomers, under a weak non-covalent bond. The monomer may be an atom, a stable molecule or a functional chemical group by itself. Dimers are characterized by rather low binding energies of the order of a few kcal/mol, with 1 kcal/mol = 0.043 eV approximately and many are more or less stable at room temperatures. Dimers are also characterized by relatively larger bond-distance. To give an analogy, a dimer is like twin cities connected by a long bridge. Of course the analogy ends there, since in reality the bridge must be strong, not weak...!

Theoretical studies on the structure of loosely bound Van der Waals dimers such as $(O_2)_2$ have been made by Bussery-Honvault and Veyret (1998). The O_4 molecule, regarded as the dimer of O_2 with parallel coplanar D_2h geometry, exhibits a rather large intermolecular separation of about 3.17 Å, as against the 'monomer' bond length $R_{0-0} = 1.21$ Å, in the O_2 molecule.

The water dimer $(H_2O)_2$ shown in the figure below, is important in many atmospheric processes.

More than two monomers can combine and form trimers, tetramers and clusters in general. Electron impact ionization is a tool for investigating dimers and clusters.

Collisions involving electrons with the water molecule have been extensively studied and an exhaustive review was compiled by Itikawa and Mason (2005). Polar molecules like water exhibit a long range anisotropic dipole potential which plays an important role in the dynamics of electron–water collisions. For example, Itikawa and Mason emphasized that the total elastic cross section 'Q_{el}' obtained in all the experimental studies is only vibrationally elastic and includes the cross section for rotational transitions, averaged over the initial rotational states and summed over the final ones. Such rotational cross sections Q_{rot} are dominant at low energies (<1 eV), important at energies below the ionization threshold but at intermediate and large electron energies the contribution of Q_{rot} towards Q_{TOT} steadily decreases and this aspect is considered in defining the grand total cross section Q_{TOT} as per equation (4.9). In the case of the H_2O molecule, the TCS Q_{rot} for rotational transition from initial rotational state $J = 0$ to final state $J' = 1$, is found to be quite large i.e., 560 $\mathring{A}^2$ at 0.12 eV, and 104.9 $\mathring{A}^2$ at 1 eV (Itikawa and Mason 2005). The magnitude of these cross sections can be gauged by comparing with the approximate geometrical cross section $\pi(R_{O-H})^2$ which comes out to be ~ 3$\mathring{A}^2$ with the molecular O–H bond length $R_{O-H} = 0.96$ Å in water molecule.

Electron–water interactions can be modelled in the form of a 4-term dynamic spherical complex potential with the absorption potential modelled with variable Δ. This was used by Joshipura et al. (2017) to calculate a complete set of TCS of e–H_2O collisions (except Q_{rot}) in the CSP-ic method. Figure 4.19 adopted from Joshipura et al. (2017) shows the various total cross sections of the water molecule. Included in the figure are the recommended data of Itikawa and Mason (2005) together with the more recent experimental data of Szmytkowski and Mozejko (2006) and Khakoo et al. (2008, 2013).

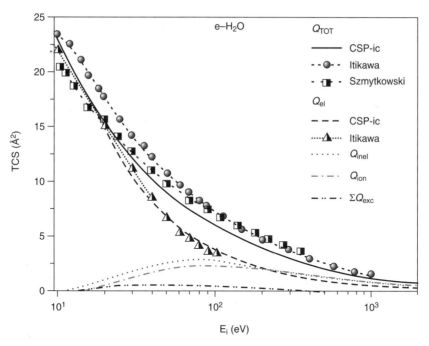

Figure 4.19 Various cross sections of e–H_2O scattering (Joshipura et al. 2017)

The Q_{TOT} derived from CSP-ic tend to fall below the recommended data up to about 200 eV while the measured data of Szmytkowski and Mozejko (2006) are even lower in the low energy range, but towards high energies the three results tend to merge. Very good agreement is found between the theoretical TCS ($Q_{el} + Q_{rot}$) and the compared values of Itikawa and Mason (2005) in Figure 4.19. Also shown in the figure are Q_{inel}, and Q_{ion}. The lowest curve displays the sum of electronic excitation TCS ΣQ_{exc} for which little data exists. Due to the permanent dipole moment, water has a large Q_{TOT} at low energy.

However, the dipole rotational contribution becomes less and less important above 100 eV. At 10 eV the elastic and the grand total cross sections practically coincide. In electron–water scattering, the inelastic peak E_P occurs at about 90 eV, while the ionization peak ε_{ion} occurs at about 100 eV.

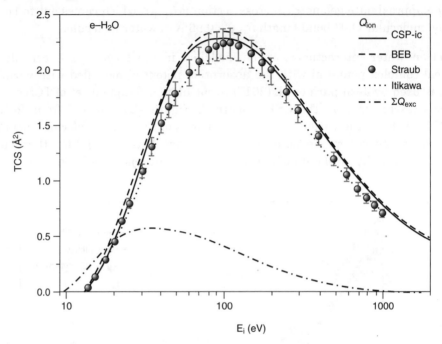

Figure 4.20 Ionization cross sections Q_{ion} (upper curves) of the water molecule compared with various other data (Joshipura et al. 2017)

Q_{ion} results are presented in a separate Figure 4.20 where CSP-ic derived cross sections are compared with the BEB data, the recommendations of Itikawa and Mason, and the measurements by Straub et al. (1998). The calculated CSP-ic ionization TCS agree quite well with the measurements of Straub et al. (1998) and also with the BEB values (NIST Database) but are a little higher than the recommended Q_{ion} of Itikawa and Mason around the peak position. The maximum value of Q_{ion} at the peak position of 100 eV is found to be 2.26 Å^2 which is about 30% of the Q_{TOT} at that energy. As the continuum channel opens at the first ionization energy, the first species to be produced is the parent ion, $(H_2O)^+$. At the Q_{ion} peak the parent ionization amounts to about 60% of the cumulative TCS Q_{ion}.

The electronic-sum ΣQ_{exc} is depicted by the lowest curve in Figure 4.20, and for this no data was recommended by Itikawa and Mason. The summed theoretical value ΣQ_{exc} equals the Q_{ion} at about 20 eV, which is slightly less than twice the ionization threshold I. The maximum value of ΣQ_{exc} occurring at about 20 eV is 0.5 Å^2 as against the recommended Q_{TOT} of almost 14 Å^2 at that energy.

In presenting the theoretical results it is necessary to give some indication of the uncertainty ascribed to the reported cross sections. In CSP-ic the input molecular parameters such as the ionization threshold, dipole moment and bond length are known accurately, therefore the major sources of error in the calculations are the potential models and the inputs of the method itself. When compared with experimental data, the Q_{ion} derived by this methodology are generally found to be within 10–12% of experimental data, while even smaller uncertainties are expected in Q_{TOT}. We therefore believe that the method will be equally valid for the other targets within the stated uncertainties.

Regarding electron scattering from OH, it may be noted that the OH radical is an important oxidizer in the terrestrial troposphere, has been detected in several astrophysical objects and is known as a source of astrophysical maser radiation. Figure 4.21, from Joshipura et al. (2017) summarises the current status of electron scattering with OH. There is only one experimental study of electron interactions with this radical, namely the electron ionization cross sections measured by Tarnovsky et al. (1996).

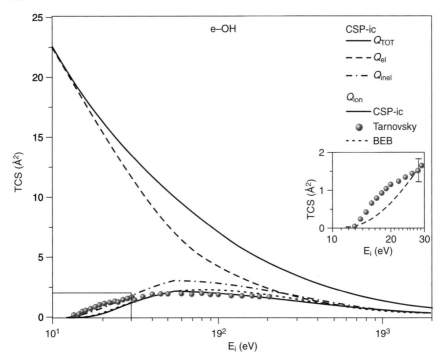

Figure 4.21 Various cross sections of electron collisions with OH radical (Joshipura et al. 2017)

Theoretical calculations (Figure 4.21) include BEB results from the quantemol-N package (Tennyson et al. 2007, Tennyson 2010) and CSP-ic calculations of Joshipura et al. (2017). CSP-ic results of Q_{TOT} shown as the top-most curve are quite large in the low energy regime due to the polar nature of the OH radical.

The Q_{el} for OH, shown by the middle curve are higher than Q_{inel} up to about 200 eV. The CSP-ic ionization TCS are in a good agreement with the experimental measurements. What is noteworthy about these ionization results is that, in the low energy regime, the measured data of Tarnovsky et al. (1996) are rising rather fast from the threshold. Up to about 30 eV the experimental data lie above the CSP-ic Q_{inel} but that is unphysical (see also inset in the figure).

4.3.2 HO_2, H_2O_2, and $(H_2O)_2$

The CSP-ic method has been extended by Joshipura et al. (2017) to investigate HO_2, H_2O_2 and $(H_2O)_2$ targets for which (to our knowledge) no experimental data exist. Each of these molecules can be considered as a two-center scattering system exhibiting a relatively larger $O - O$ bond length (Table 4.5). For a molecule XY having a relatively large bond length denoted as R_d (R_{OO} presently) between two constituents X and Y, i.e., atoms or functional chemical groups, the simple approximation known as the Additivity Rule (4.7) may be used. The AR neglects the overlap or screening of charge-clouds of constituents X and Y and therefore needs appropriate corrections, which can be included in two different ways. In an elementary screening correction to be called *sc1*, the static or geometrical overlap of X and Y in the molecule XY having their centres separated by R_d, is considered. Using the exact geometrical formula of two intersecting circles, given in (Weisstein, Wolfram Web Resource) the (maximum) overlap area between X and Y is derived. If A_x and A_y are the static or geometrical areas of X and Y, then an overlap fraction is defined by

$$F_{ov} = \frac{[(A_x + A_y) - A_{ov}]}{(A_x + A_y)} \qquad (4.10)$$

Thus for a cross section 'Q' the AR corrected in *sc1*, can be written as (Joshipura et al. 2017),

$$Q_{sc1}(XY) = F_{ov}[Q(X) + Q(Y)] \qquad (4.11)$$

An overlap corresponding to polarizability volume can also be introduced in a similar way if the polarizability data are known accurately.

A dynamic or energy dependent screening correction to AR for electron–molecule scattering cross sections was introduced by Blanco and Garcia (2003), see also Chiari et al. (2013). Joshipura et al. (2017) employed this screening procedure to functional groups and not to the individual atoms (Blanco and Garcia 2003) of the molecule. Thus in the dynamic correction *sc2*, the calculated TCS $Q(X)$ and $Q(Y)$ are each multiplied by a corresponding screening factor, and a screening-corrected AR as in Blanco and Garcia (2003) is expressed as,

$$Q_{sc2}(XY) = S_x Q(X) + S_y Q(Y) \qquad (4.12)$$

where, the screening factors are

$$S_x = 1 - \frac{1}{2}\left[\frac{Q(y)}{4\pi R_d^2}\right]; \; S_y = 1 - \frac{1}{2}\left[\frac{Q(x)}{4\pi R_d^2}\right] \tag{4.13}$$

Consider Figure 4.22 from Joshipura et al., representing e–HO$_2$ scattering. Here, the symbol Q_{TOT} denotes the grand TCS in AR derived without any screening correction, while Q_{TOT1} and Q_{TOT2} correspond to corrections *sc1* and *sc2* as defined in equations (4.11) and (4.12, 4.13) respectively. Similar notations are adopted for the elastic and the ionization TCS. All the dipole rotational cross sections Q_{rot} are calculated at the direct target level. Also shown in Figure 4.22 are the results of an earlier BEB calculation (NIST database). The AR results are expected to overestimate the cross section Q_{TOT} below 300 eV. The correction *sc1* produces a small reduction in Q_{TOT} as well as in Q_{el} (rotationally inclusive). At a low energy of 20 eV, both Q_{TOT1} and Q_{el1} (Figure 4.22) are lower than the Q_{TOT} and Q_{el} by about 7%. The AR with *sc2* results in an energy dependent reduction in all three TCS. Once again at 20 eV the effect of *sc2* is to decrease the Q_{TOT} and Q_{el} by 14% and 12% respectively; however, the *sc2* correction becomes insignificant above 300 eV.

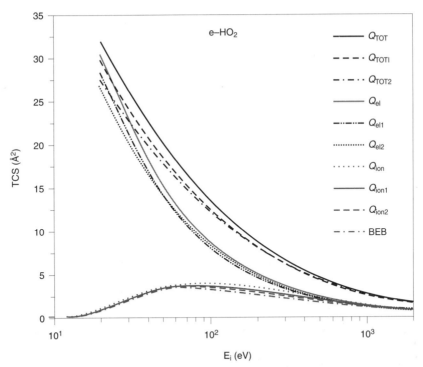

Figure 4.22 Calculated total cross sections of HO$_2$ (Joshipura et al. 2017); BEB Q_{ion} (NIST database)

The lowest set of curves in this figure presents Q_{ion} for HO$_2$. The effect of *sc1* on Q_{ion} persists over the entire energy range while that of *sc2* is appreciable only in the peak region. At the peak, the *sc2* reduces the AR Q_{ion} by about 4% as against (about) a uniform 10% reduction observed with *sc1*. For this exotic radical species, the maximum Q_{ion} predicted in BEB is 3.70 Å^2 quite close to the *sc2* value of 3.35 Å^2. Joshipura et al. ascribed an uncertainty of about 10% to their derived values.

Figure 4.23 displays electron impact studies on H$_2$O$_2$ molecule. Similar trends are seen as in HO$_2$. However at 20 eV the *sc1* reduces both Q_{TOT} and Q_{el} by about 5% while the dynamic screening *sc2* reduces both these TCS by about 23%. The effect of *sc2* on cross sections Q_{TOT} and Q_{el} diminishes above 100 eV. In Q_{ion} the decrease due to *sc1* and *sc2* is 6% and 4% respectively at the peak position.

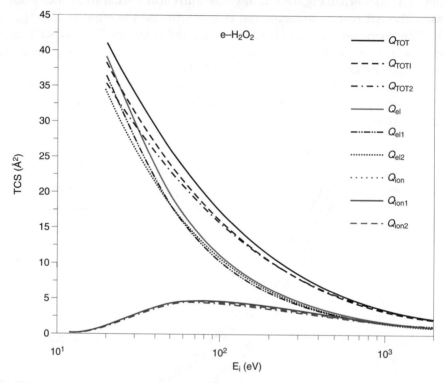

Figure 4.23   Theoretical TCS for electron scattering from the H$_2$O$_2$ molecule (Joshipura et al. 2017)

Finally we turn to the water dimer (see also Box 4.2) for which low energy calculations up to 10 eV have been reported by Bouchiha et al. (2008) and Caprasecca et al. (2009). Figure 4.24 shows the CSP-ic results on (H$_2$O)$_2$ estimated using the additivity rule along with screening corrections. An equilibrium geometry (EQ) is used for the dimer. At 10 eV the total elastic cross section of Caprasecca et al. (2009) is close to 19.36 Å^2 whereas the calculated rotational cross section Q_{rot} (J = 0 → 1) turns out to be higher at 25.9 Å^2. Once again these cross sections are large due to the large dipole moment of the dimer.

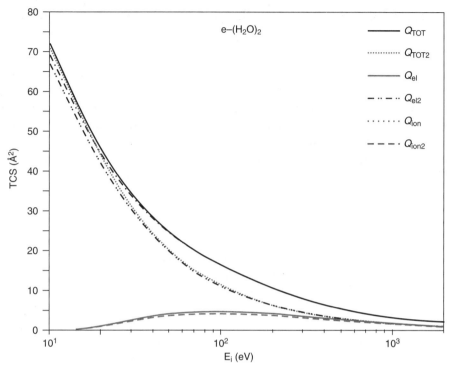

Figure 4.24 Theoretical TCS for electron scattering from the water dimer (Joshipura et al. 2017)

In the case of the water dimer the effect of the geometrical correction *sc1* is small (negligible), due to rather large separation between the two monomers. The correction *sc2* reduces the Q_{TOT} by about 3% and Q_{el} by about 4% at 20 eV and at the peak of the ionization the AR cross section is reduced by just 1%.

In Table 4.6 the maximum ionization cross sections and the corresponding electron energies (ε_{ion}) are displayed for the five targets. For the two-center targets the screening correction *sc2* has been employed for this table.

Table 4.6 Maximum Q_{ion} and corresponding peak position (Joshipura et al. 2017)

Target	Peak position of Q_{ion} (eV)	Maximum Q_{ion} ($Å^2$)
H_2O	90	2.26
OH	70	2.13
HO_2	70	3.45
H_2O_2	70	4.01
$(H_2O)_2$	90	4.49

4.4   METHANE, CH₄, AND THE RADICALS CHₓ (x = 1, 2, AND 3)

4.4.1   CH₄

Methane is one of the simplest, symmetric hydrocarbons and has attracted considerable attention due to its high global warming potential. Properties of this molecule are given in Table 1.1, Chapter 1. Song et al. (2015) have recently published a comprehensive data review of electron scattering from methane updating the earlier review of Karwasz et al. (2001). An almost isotropic electron interaction with methane makes it an ideal target for the approximate theoretical method CSP-ic (Joshipura et al. 2004, Vinodkumar et al. 2006).

Figure 4.25 summarises electron scattering data for methane. The Q_T theoretical data of Joshipura et al. (2004) are consistent with the measurements of Zecca et al (2001). More than two decades ago Jain and Baluja (1992) performed complex potential calculations to determine the total (elastic *plus* inelastic) cross sections for a large number of diatomic and polyatomic molecules including CH_4. Their data, included in Figure 4.25, appear to underestimate Q_T. The lower curves in this figure represent Q_{inel}, Q_{ion} and ΣQ_{exc} respectively. The CSP-ic ionization cross sections match well with the measured values of Chatham et al. (1984) as well as Nishimura and Tawara (1994) at least within their experimental uncertainties. The quoted CSP-ic Q_{ion} are also in agreement with the recommended data of Song et al. (2015).

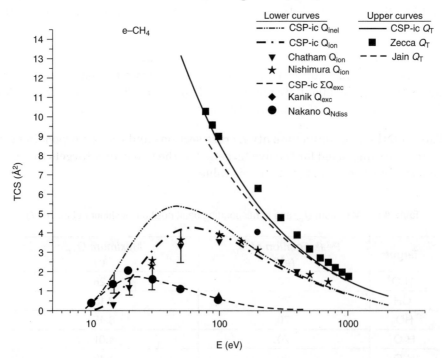

Figure 4.25   CSP-ic results along with comparisons on e–CH_4 scattering from Joshipura et al. (2004)

Table 4.7 compares the CSP-ic values with those of Song et al. (2015) at 100 eV.

Table 4.7 TCS (Å^2) of methane at 100 eV incident energy; CSP-ic of Joshipura et al. (2004) are compared with the recommended data of Song et al. (2015); *the Q_{inel} of Song et al. are calculated from their ($Q_T - Q_{el}$)

Q_T		Q_{el}		Q_{inel}	
CSP-ic	*Recommended Data*	*CSP-ic*	*Recommended Data*	*CSP-ic*	*Recommended Data**
9.60	9.24	5.06	(4.09 ± 1.48)	4.54	5.14

Since all of the electronic states of methane are purely dissociative in nature, ΣQ_{exc} also represents the total neutral dissociation cross section Q_{NDiss}. Methane dissociates into neutral fragments upon electronic excitation, as follows

$$e + CH_4 \rightarrow CH_3 + H,$$

$$\rightarrow CH_2 + H_2$$

$$\rightarrow CH + 3H \text{ or } H_2 + H \tag{4.14}$$

More than two decades ago, Nakano et al. (1991) measured total cross sections for the neutral dissociation of methane into CH_3 and CH_2. The sum of the measured TCS of Nakano et al. denoted by Q_{NDiss} is exhibited in Figure 4.25. It is interesting to note that there is good agreement between the summed data of Nakano et al. and the CSP-ic values of summed total excitation cross sections ΣQ_{exc}, and both predict the peak region to be around 25 eV. However, it must be noted that high uncertainties are expected in the above measurements. The semi-empirical total electronic excitation cross section as given by Kanik et al. (1992) are seen to be slightly higher than the data of Nakano et al. (Figure 4.25).

The Q_{inel} for CH_4 reaches a peak around 50 eV, at which energy the CSP-ic ratio R_p is quite close to 70%. Ionization cross section Q_{ion} reaches its maximum at about 70 eV. A bar chart showing the relative importance of different scattering channels in CH_4 at this energy is given in Figure 4.26.

Special attention should be given to the calculations of total elastic cross sections for CH_4 over a wide energy range from 0.01 eV to 2 keV, as performed by Vinodkumar et al. (2011). The significance of that work lies in the fact that it was an attempt to smoothly combine two different theories viz., the SCOP (Spherical Complex Optical Potential) and the R-matrix (in quantemol-N) method. Therefore, this approach may be called a hybrid model. The SCOP method begins to break down near the first ionization threshold and does not hold at lower energies, where one needs accurate theories like the R-matrix method. It is desirable to have a smooth transition at the cross over from the R-matrix (quantemol-N) results to SCOP values, and this can be seen in

Figure 4.27, showing the total elastic cross sections of methane over a large energy range, as derived by Vinodkumar et al. (2011).

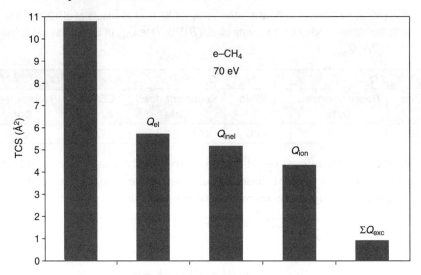

Figure 4.26   A bar-chart showing TCS of e–CH$_4$ scattering at 70 eV (Joshipura et al. 2004); cross section Q_T is shown by the first bar; Q_{el} and Q_{inel} are 52.5% and 47.5% of Q_T respectively; Q_{ion} and ΣQ_{exc} are 83% and 17% of Q$_{inel}$ respectively

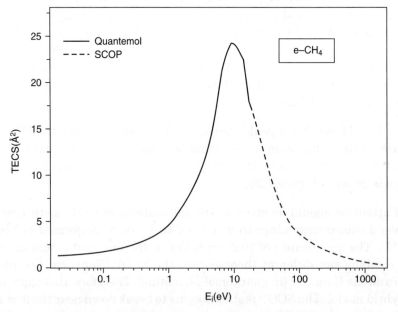

Figure 4.27   Total elastic cross sections for electron–methane scattering from very low to high energies; continuous black curve: quantemol-N results; and dashed curve: SCOP results, merging near the ionization threshold ~15 eV, Vinodkumar et al. (2011)

4.4.2   Radicals: CH_x (x = 1, 2 and 3)

Reactive radicals CH_x (x = 1, 2 and 3) are produced in neutral dissociation of the parent molecule CH_4. Radicals in general, are physically stable but chemically reactive atomic/molecular species. Owing to their electronic configurations, they are highly ionizable and polarizable as well.

Due to the difficulty in preparing radical targets for electron scattering experiments there are few, if any, experimental studies of electron scattering from radicals such as CH_x; indeed there are perhaps no measurements and only one calculation for Q_T of CH_x. Figure 4.28 shows electron scattering with CH (methylidyne) radical, where the theoretical Q_T as well as Q_{ion} calculated in CSP-ic by Vinodkumar et al. (2006) are shown. The grand total cross sections of CH are large especially at lower energies due to the dipole potential. The measured ionization data from Tarnovsky et al. (1996), the BEB Q_{ion} values from the NIST database and CSP-ic ionization data derived from Vinodkumar et al. are all in general agreement particularly when experimental uncertainties are allowed for.

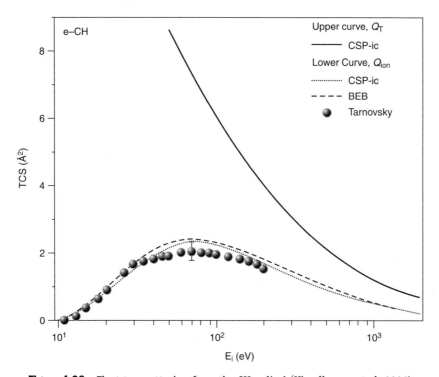

Figure 4.28   Electron scattering from the CH radical (Vinodkumar et al. 2006)

Electron scattering from the CH_2 radical is shown in Figure 4.29 (Vinodkumar et al. 2006). The Q_{ion} data are the BEB results, the CSP-ic data of Vinodkumar et al. and the measured data from two groups viz., Tarnovsky et al. (1996) and Baiocchi et al. (1984).

Once again the Q_{ion} are all in general agreement particularly when experimental uncertainties are allowed for.

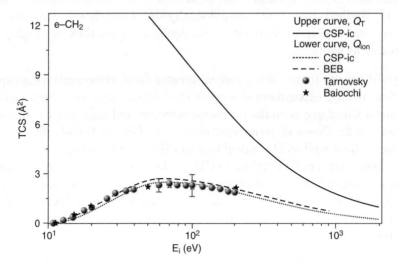

Figure 4.29 Electron scattering from the CH_2 radical (Vinodkumar et al. 2006)

The total and the ionization TCS of methyl radical CH_3 are plotted in Figure 4.30, in which comparisons of the Q_{ion} are made with both experimental and theoretical results. Once again the BEB and CSP-ic calculations are in good agreement with the experimental results. Finally, it is also observed that in the region of maximum values, the ionization cross sections increase slightly along the sequence $CH–CH_2–CH_4–CH_3$ (Vinodkumar et al. 2006). This is understood in terms of the respective molecular properties.

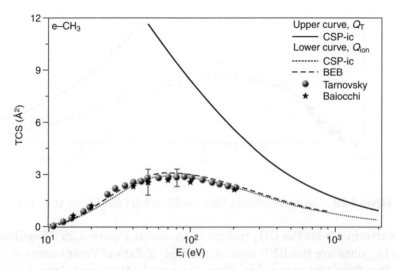

Figure 4.30 Electron scattering from the CH_3 radical (Vinodkumar et al. 2006)

4.5 OTHER COMMON MOLECULES AND THEIR RADICALS

4.5.1 The ammonia molecule and the radicals NH_x (x = 1, 2)

NH_3

The pyramidal molecule NH_3 is endowed with a dipole moment of 0.58 au and a relatively small polarizability $\alpha_0 = 2.26\ \text{Å}^3$. Ammonia is widely used in industrial processes and is a natural biogas. It is also widely used in industrial plasmas and is often a product in nitrogen rich plasmas where electron induced chemistry is prevalent. Earlier studies of electron ammonia scattering were reviewed by Karwasz et al. (2001).

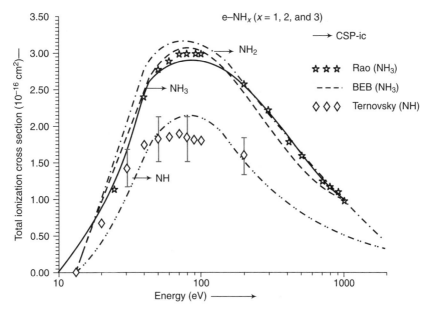

Figure 4.31 Electron impact ionization cross sections for the NH_3 molecule and its radicals; NH_3, CSP-ic *continuous curve*: Joshipura et al. (2001); *dashes*: BEB (NIST database); *stars*: experimental data Rao and Srivastava (1992); NH_2 *topmost chain curve* and NH *lowest chain curve*: Joshipura et al. (2001); NH experimental data *diamonds* Tarnovsky et al. (1997)

Figure 4.31 displays the Q_{ion} of the parent molecule NH_3 and the two relevant radicals. A BEB derivation of Q_{ion} was reported by Hwang et al. (1996) and the CSP-ic method, in its earlier version was reported by Joshipura et al. (2001) and more recently by Vinodkumar et al. (2010) with an initial input $R_p = 0.67$. A recent fixed-nuclei complex potential calculation on e–NH_3, PH_3 scattering in the energy range 0.1–100 eV has been reported by Kaur et al. (2015). The results of Kaur et al. included elastic differential as well as integral cross sections and total (elastic *plus* inelastic) cross sections at intermediate energies, and comparisons with measurements were made by them with available cross sections. However the numerical values of their inelastic cross section Q_{inel} were not available, and hence could not be incorporated in our discussion on NH_3.

Amino radicals: NH_x (x = 1, 2)

The radicals NH and NH_2 (called amino radicals) are produced in $N_2 - H_2$ plasmas. It is interesting to compare the electron ionization of the radicals NH_x (x = 1, 2) with respect to parent molecule NH_3. Referring to Figure 4.31, one finds that the CSP-ic Q_{ion} (continuous curve) for the ammonia molecule are consistent with the BEB values (NIST database) and with the measurements of Rao and Srivastava (1992). For the radicals NH and NH_2, ionization measurements were carried out by Tarnovsky et al. (1997). The measured data for the NH radical (Tarnovsky et al. 1997) rise rather rapidly and exhibit a flat looking peak region. The CSP-ic Q_{ion} values are within the experimental uncertainties (Figure 4.31) around the peak region. Amongst the three targets, the NH radical exhibits the lowest peak of Q_{ion}. The differences in the $NH–NH_2–NH_3$ sequence can be interpreted in terms of the respective ionization energies as well as the bond lengths.

4.5.2 Sulphur molecules: SO, SO_2, and H_2S

Amongst a host of sulphur compounds we select here the diatomic radical SO along with its parent molecule SO_2. The S–O bond lengths and dipole moments of SO and SO_2 are nearly equal. Calculations of the electron scattering cross sections for these two targets, based on CSP-ic were made by Joshipura et al. (2008). Figure 4.32 shows the TCS of e–SO scattering. Note that, unlike in the previous cases, the uppermost curve in this figure shows the total elastic cross sections Q_{el} (without Q_{rot}), which tend to fall below the Q_{inel} beyond 70 eV. The point-dipole rotational excitation TCS Q_{rot} are also shown here by the chain curve on the left. CSP-ic derived ionization TCS are compared in Figure 4.32 with the measured data of Tarnovsky et al. (1995). The dashed curve in this figure is the Q_{ion} using the AR also from Tarnovsky et al. The agreement is fairly good, up to the peak region, but thereafter the measured data falls below the calculated data.

A full set of TCS of e–SO_2 collisions is plotted in Figure 4.33. The uppermost curve depicts the grand total cross section Q_{TOT} calculated by Joshipura et al. (2008) obtained by adding Q_{rot} to Q_T, compared with the measurements of Szmytkowski and Maciag (1986), and Zecca et al. (1995). Joshipura et al. (2008) tend to overestimate low energy Q_{TOT} when compared with the measured data of Szmytkowski and Maciag. The calculated Q_{el} are compared with the theoretical data of Raj and Tomar (1997) and the integrated elastic TCS of Orient et al. (1982); there is a partial agreement amongst these results. The CSP-ic Q_{ion} for SO_2 molecule are in a fair agreement with BEB theory (NIST database) as well as the experimental data of Cadez et al. (1983) and Basner et al. (1995) but the measured ionization data of Orient and Srivastava (1984) are notably higher, even allowing for experimental error bars.

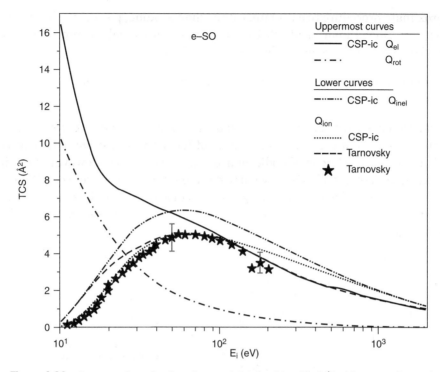

Figure 4.32 Cross sections Q_{el}, Q_{rot}, Q_{inel}, and Q_{ion} for SO radical (Joshipura et al. 2008)

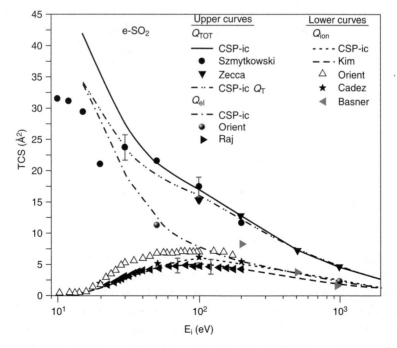

Figure 4.33 Electron scattering from the SO_2 molecule (Joshipura et al. 2008)

An approximate procedure to estimate, rather than assume, parameter 'R$_p$' for a target was given by Vinodkumar et al. (2010, 2011) in a method called Improved CSP-ic or ICSP-ic. The method however needs further development.

Let us now turn to angular polar H$_2$S molecule and focus on ionizing collisions of electrons.

Several datasets exist for Q_{ion} but, as shown in Figure 4.34 there appear to be two different clusters of data. The measured data of Rao and Srivastava (1993) agree with the BEB data and ICSP-ic of Vinodkumar et al. (2010, 2011). However, experiments of Lindsay et al. (2002) and the very early data of Otvos and Stevenson (1956) favour higher cross sections. The top curve is Q_{inel} from Vinodkumar et al. (2011) which is still consistent with both clusters of experimental data. These discrepancies require that Q_{ion} be re-measured.

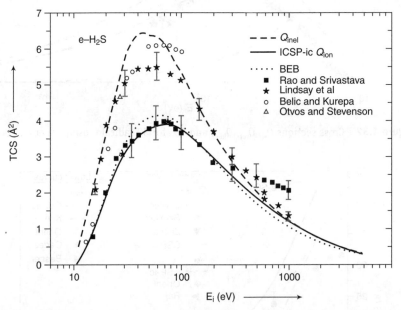

Figure 4.34 Total ionization cross sections of H$_2$S molecule – Vinodkumar et al. (2011); from top down, Q_{inel}, *dashed curve*: CSP-ic; *continuous curve*: ICSP-ic, BEB (NIST database); *open circle*: measurements by Belic and Kurepa (1985); *star*: Lindsay et al. (2002); *filled squares*: Rao and Srivastava (1993), *open triangles*: Otvos and Stevenson (1956)

Finally in this sub-section a brief mention about the disulphur molecule S$_2$ which is a simple but rare sulphur species, having a rather large bond length of 1.889 Å and ionization threshold of 9.356eV. S$_2$ has been determined to be a minor constituent of Jupiter's satellite Io where electron collisions are known to occur. BEB data on Q_{ion} (NIST data Base) indicate a peak value of 8.1 Å^2 around 60 eV, while in Vinodkumar et al. (2010) the peak Q_{ion} of S$_2$ is 5.85Å^2. Total (complete) cross sections of electron impact on S$_2$ molecules were calculated by Naghma et al. (2014) and benchmarked to experiments on O$_2$.

Electron impact ionization cross sections for several other sulphur containing molecules such as CS, CS_2, OCS were investigated in the CSP-ic formalism by Vinodkumar et al. (2010).

4.6 REACTIVE SPECIES CN, C_2N_2, HCN, AND HNC; BF

4.6.1 The Cyano radical and Cyanogen

The cyano radical CN was amongst the first radicals detected in interstellar medium together with cyanogen C_2N_2 with N≡C–C≡N geometry. Hydrogen cyanide HCN and its isomer hydrogen isocyanide HNC have been found in the cold interstellar medium (ISM) where the concentration of HNC often exceeds that of HCN (Hirota et al. 1998). Recently these compounds have been identified in the atmosphere of Titan where they are important in the formation of a dense haze arising from CN containing dust. All the five listed targets are polar molecules. Electron impact ionization of these exotic molecules was carried out in CSP-ic by Pandya et al. (2012) who also made a comparative study of Q_{ion} with the 14-electron systems N_2, CO, HCN, HNC, C_2H_2, and BF.

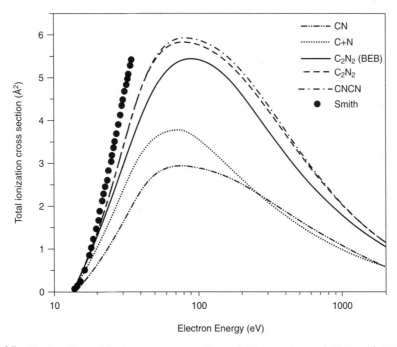

Figure 4.35 Electron impact ionization cross sections of CN (lowest curve), C_2N_2 and CNCN (upper most curve); CSP-ic and AR (Pandya et al. 2012) compared with experimental results of Smith (1983) for C_2N_2 and BEB calculation for C_2N_2.

Figure 4.35 adopted from Pandya et al. exhibits the Q_{ion} for CN and C_2N_2, in which the curve denoted by (C + N), stands for Q_{ion} of CN in simple AR. In the case of C_2N_2, the calculated Q_{ion} is twice the CSP-ic Q_{ion} (CN), which is derived at the ionization threshold of C_2N_2. At the peak position (80 eV) the AR Q_{ion} of C_2N_2 reduces by about 6% with the screening correction *sc2* (not shown here) discussed above. Experimental results of Smith (1983) for C_2N_2 are limited to low energy below the ionization peak, and these data tend to overestimate the cross section as can be seen in Figure 4.35. The Q_{ion} for isocyanogen CNCN is also included in Figure 4.35. In view of the interesting geometry of CNCN isomer ($CN_1 - CN_2$) its total ionization cross sections obtained in CSP-ic theory are also shown in this figure.

4.6.2 HCN and HNC; BF

Figure 4.36 exhibits the theoretical Q_{ion} for HCN and HNC, along with the BEB Q_{ion} obtained from the quantemol – N package mentioned earlier.

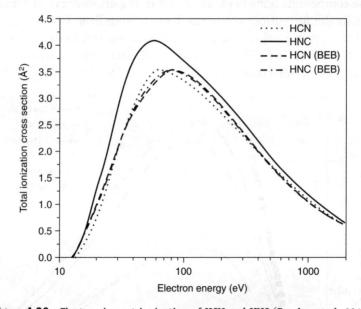

Figure 4.36   Electron impact ionization of HCN and HNC (Pandya et al. 2012)

The HNC isomer exhibits higher Q_{ion} than HCN up to the peak. Discrepancies in HNC observed by Pandya et al. (2012) with respect to the BEB values seem to be due to the difference in the value of 'T' employed. The CSP-ic work on HNC employed the experimental value $I = 12.50$ eV, as against the value $I = 13.29$ eV – the Koopman theorem value which was employed in BEB.

The present results on boron monofluoride BF, shown in Figure 4.37, are obtained by single-center charge density employed to construct the total complex potential. The CSP-ic results for BF are of interest due to lack of any experimental data. The Q_{ion} are calculated using two models of V_{abs} with the energy parameter Δ varying over a small range and with $\Delta = I$ for BF (Figure 4.37). In this Q_{ion} plot, the symbol (B + F) in the figure caption corresponds to simple AR, while the BEB data for this molecule is generated from the Quantemol – N package (Tennyson et al. 2007, Tennyson 2010).

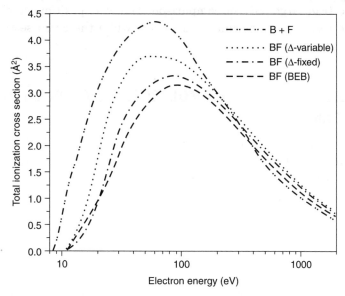

Figure 4.37 Ionization cross sections for BF (Pandya et al. 2012)

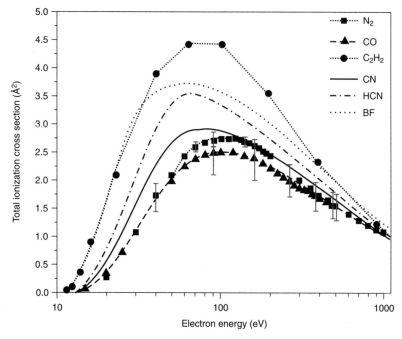

Figure 4.38 Q_{ion} of five iso-electronic molecular systems, and a 13-electron target CN (Pandya et al. 2012); *square*: measurements of N_2 (Straub et al. 1996); *triangle*: measurements of CO (Orient and Srivastava 1987); *circle*: measurements of C_2H_2 (Hayashi et al. 1990); CSP-ic results; continuous *curve*: for CN; dash dot: for HCN; *dot*: for BF

Now, consider Figure 4.38 showing an interesting comparison of trends in ionization cross sections amongst the five 14 – electron systems N_2, CO, C_2H_2, HCN, and BF, along with the 13-electron target CN for comparison. The peak positions and magnitudes of Q_{ion} reflect the respective molecular properties, while the cross sections tend to be similar at high energies.

4.7 METASTABLE SPECIES OF MOLECULAR HYDROGEN AND NITROGEN

As discussed in Chapter 3 several molecules have long-lived excited metastable states which can strongly influence the chemistry of their local medium. We confine the discussion here to the electronically excited metastable states (EMS) of two of the most well-known molecules H_2 and N_2. Electron collision studies with metastable states are rather scarce (see also Box 3.1, Chapter 3) and for these two targets there are no scattering experiments with which to compare theoretical evaluations.

4.7.1 $H_2^*(c^3\Pi_u)$

CSP-ic calculations on electron scattering and ionization of the metastable H_2^* in the electronic state $c^3\Pi_u$ (radiative life time ~1 ms) were reported by Joshipura et al. (2010). Table 4.8 displays the properties of the H_2 ground state(GS) and the metastable H_2 species. For the excited metastable electronic state $c^3\Pi_u$ of H_2, the experimental ionization threshold is I = 3.66 eV according to the NIST Chemistry WebBook (2017), and a theoretical value 2.82 eV was given by Branchett et al. (1990). $H_2(c^3\Pi_u)$ eventually dissociates into atomic species H(1s) and H(2p).

Table 4.8 Properties of H_2 (GS) and H_2^* (Joshipura et al. 2010), [+]Branchett et al. (1990)

Target Species	First ionization energy eV	Bond length in Å	Polarizability α_0 in Å^3	Peak position E_p in eV
H_2 (GS)	15.42	0.74	0.80	50
$H_2^*(c^3\Pi_u)$	3.66	1.03	2.19	10
–	2.82[+]	–	–	10

Let us consider the interactions of incident electrons with molecular hydrogen initially in the EMS viz., $H_2^*(c^3\Pi_u)$. The calculation proceeds by constructing an appropriate single-center electron charge density and hence deriving the total complex potential. The Q_{ion} results based on CSP-ic (Joshipura et al. 2010) are plotted graphically in Figure 4.39 in which a comparison with H_2 (GS) has also been made. Though the TCS Q_T is not shown here, it was found by Joshipura et al. that at 10 eV, the Q_T of H_2^* is larger than that of H_2 (GS) by a factor of 3.

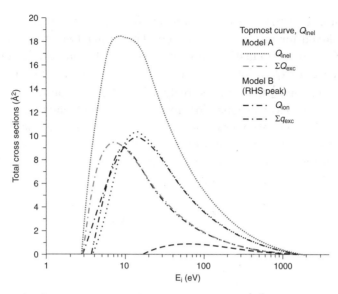

Figure 4.39 Ionization cross sections of electron impact on $H_2^*(c^3\Pi_u)$, Joshipura et al. (2010) lowest dashed curve, H_2 ground-state

The ionization TCS of the unknown target H_2^* were calculated by Joshipura et al. (2010) in two ways. They defined model A in which the energy parameter Δ (of the absorption potential V_{abs}) was chosen as varying smoothly from 2.82 eV to 3.66 eV, while in model B it was fixed at 3.66 eV. At the peak position (Figure 4.39), the Q_{ion} of H_2^* is about ten times larger than that of H_2 (GS), and the peak is also shifted to lower energy accordingly.

4.7.2 $N_2^*(A^3\Sigma_u^+)$

Finally consider electron scattering from metastable molecular nitrogen. There are three metastable states of N_2 molecule and these are $A^3\Sigma_u^+$, $a^1\Pi_g$ and $E^3\Sigma_g^+$ situated at 6.17, 8.55 and 11.87 eV above the ground electronic state respectively. Electron scattering from the lowest excited metastable state $A^3\Sigma_u^+$, which has lifetime close to 2 seconds, was studied in the CSP-ic calculations of Joshipura et al. (2009). The properties of the N_2 molecule in both its electronic ground state and in the EMS $A^3\Sigma_u^+$ are compared in Table 4.9.

Table 4.9 Properties of N_2 (GS) and N_2^*, see Joshipura et al. (2009)

Target Species	First ionization energy eV	Bond length in Å	Polarizability α_0 in Å^3	Peak Q_{inel} position E_p in eV
N_2 (GS)	15.58	1.098	1.71	75
$N_2^*(A^3\Sigma_u^+)$	9.52	1.29	2.87	55
	10.10	–	–	55

In Figure 4.40, the CSP-ic cross sections of $N_2^*(A^3\Sigma_u^+)$ obtained by Joshipura et al. (2009) are displayed for mutual comparison. The upper two curves are for the total cross sections Q_T and total elastic cross sections Q_{el}, whereas the lower curves are for inelastic cross sections Q_{inel}, total ionization cross sections Q_{ion} and the summed-total excitation cross sections ΣQ_{exc}, respectively. Since N_2^* is a relatively unknown target for electron scattering calculations, Joshipura et al. (2009) considered two models for the calculation of the ionization cross section Q_{ion}. In model A, Q_{ion} is calculated for N_2^* in the usual CSP-ic method by assuming $R_p = 0.70$. In model B the ratio at the inelastic peak, i.e., R_p is chosen to be 0.60 and the ionization potential of N_2^* is the appearance potential 10.10 eV (Armentrout et al. 1981).

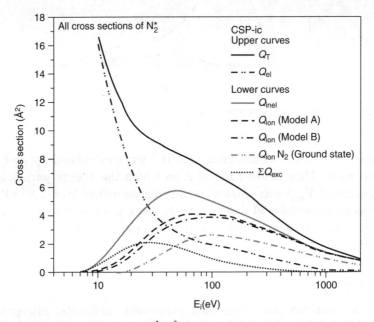

Figure 4.40 TCS for electron impact on N_2^* $(A^3\Sigma_u^+)$, Q_{ion} of N_2 (GS) also included; Joshipura et al. (2009)

The calculated ionization cross sections for N_2^* are higher than those of the N_2 (GS), in view of the lower ionization threshold and a larger bond length of the EMS compared to the GS. The ratio of the maximum Q_{ion} values of N_2^* and N_2 is 1.59 in model A, and 1.50 in model B.

4.8 GENERAL TRENDS AND CORRELATIONS

Some generalization or correlation of the electron impact cross sections with basic molecular properties is examined as we close this chapter, particularly regarding maximum ionization cross sections σ_{max}.

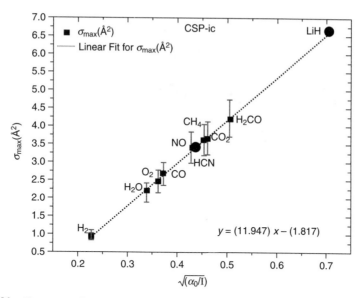

Figure 4.41 Linear correlation of σ_{max} as a function of $\sqrt{(\alpha_0/I)}$ for a variety of ground-state molecular targets

Similar to the previous chapter, we consider here the general trends or correlations amongst electron–molecule cross sections, by defining a target–property parameter; a target–property parameter $P_A = \sqrt{(\alpha_0/I)}$, where α_0 is the average static dipole polarizability of the molecule. As shown in Figure 4.41, this quantity correlates very well with the peak ionization cross section denoted by σ_{max}. The error bars indicate 10% uncertainty in σ_{max}. The quality of the fit is excellent (with $R^2 = 0.99$). However, it must be noted that the correlation is demonstrated for only a few of the smaller molecules.

✿ ✿ ✿

5

ELECTRON MOLECULE SCATTERING AND IONIZATION – II

Other Polyatomic Molecules and Radicals

This chapter covers larger polyatomic targets composed of four or more constituent atoms and, where appropriate, the radicals formed from such molecules in electron impact dissociation. No exhaustive coverage of polyatomic molecules is claimed here; instead the focus is on representative polyatomic targets of chemical families. For each target, comparisons have been made between theoretical and experimental data, together with commentary on the different electron collision processes. A list of the molecules reviewed in this chapter is given in Box 5.1. In developing a theoretical approach for the scattering of electrons from such targets if the molecule has a single centre with respect to which the molecular electron charge density can be expressed to construct the total complex potential, then the scattering formalism is greatly simplified. However, with an increase in the number of constituent atoms two-centre systems are apparent, especially in the small hydrocarbons and corresponding fluorocarbons. In such cases the group additivity rule (group-AR) may be applied, for example in molecules such as C_2H_4 or C_2H_6 where the C – C bond length is larger than the C – H bond length. However, this is only a partial solution and the neglect of overlap or screening in the molecule when the Additivity Rule is applied, is compensated to an extent, by considering screening corrections as discussed in the previous chapter. Joshipura et al. (2017) have provided further details.

The semi-empirical CSP-ic methodology has been applied to all these targets. As discussed earlier, this approach starts with the partitioning of the total inelastic cross section Q_{inel} of a molecule into two cumulative quantities: ΣQ_{exc}, corresponding to accessible electronic excitations and, ΣQ_{ion}, corresponding to accessible ionizations. The cumulative ionization cross section, denoted simply by Q_{ion}, includes single, double, direct, dissociative and other ionization processes depending on the incident electron energy. Relative partitioning between electronic excitation and ionization is characterized by two parameters, R_p at peak energy E_p and R' at a higher energy. With approximate initial input $R_p = 0.7$, the R' value is generated iteratively, as explained in the previous chapters. We recall that the continuous energy dependence of the ratio function $R(E_i)$ requires parameters a, C_1 and C_2. The relevant expression is reproduced below

$$R(E_i) = 1 - C_1 \left[\frac{C_2}{U+a} + \frac{\ln U}{U} \right] \tag{5.1}$$

BOX 5.1: MOLECULES REVIEWED IN THIS CHAPTER

NF_3 and radicals NF_x (x = 1, 2)

CF_4 and radicals CF_x (x = 1,2 & 3)

H_2CO and CHO, $HCOOH$

C_2H_2, C_2H_4, C_2H_6

C_2F_2, C_2F_4, C_2F_6

CH_3OH, CH_3I, CF_3I

SiH_4, SiF_4, GeH_4, GeF_4; relevant radicals

CCl_4, $SiCl_4$, and radicals $SiCl_x$ (x = 1, 2 & 3)

Si_2H_6, Ge_2H_6; $Si(CH_3)_4$

SO_2XY (X, Y = F, Cl), SF_6, H_3PO_4

C_3H_4, C_3H_6, C_3H_8

C_4H_6, C_4H_8, c-C_4H_8

C_4F_6 and c-C_4F_8

C_4H_4O, C_4H_8O, C_6H_8O

5.1 SMALL POLYATOMIC MOLECULES

5.1.1 NF_3 and radicals NF_x (x = 1, 2)

Nitrogen trifluoride was first synthesized about a hundred years ago and has since been detected in naturally occurring rocks (Harnisch et al. 2000). It is used by the semiconductor industry for plasma etching and cleaning of semiconductors (Chow and Steckl 1982, Donnelly et al. 1984) for which dissociative ionization is an important phenomenon to quantify. Dissociative ionization of small molecules like NF_3 is also of interest in testing the quantum mechanical constraints on the distribution of different fragments in ionization processes (Jiao et al. 1999).

The molecular properties of NF_3 are, $R_{N\text{-}F}$ = 1.365 Å, I = 12.94 eV and average polarizability α_0 = 2.81Å^3 (see CCCBDB database). A few experimental investigations to measure cross sections for electron ionization of NF_3 have been reported in the literature. Tarnovsky et al. (1994) investigated e–NF_3 scattering and measured absolute cross sections for the formation of NF_x^+ (x = 1–3), and other ions. Haaland et al. (2001) determined absolute cross sections for the formation of N^+, F^+, NF_x^+ (x = 1–3) and the dications NF_x^{2+} (x = 1–3). NF_3 is one of those targets for which an unusually large discrepancy is observed between experiments and theories (Deutsch et al. 1994, Szmytkowski et al. 2004). These discrepancies led to a joint experimental and theoretical investigation carried out by two groups in India, as reported by Rahman et al. (2012), wherein e–NF$_3$ scattering was examined theoretically in the SCOP and CSP-ic methodology. Total ionization cross sections Q_{ion} (included in figures 5.1 and 5.2 reproduced from Rahman et al. 2012) were measured directly by Krishnakumar group in India.

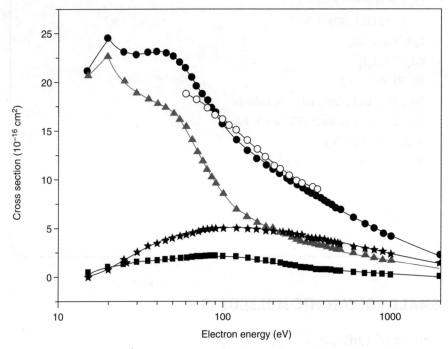

Figure 5.1   Calculated TCS of NF$_3$ from Rahman et al. (2012), *filled circles*: Q_T in CSP-ic; *open circles*: measured Q_T data of Szmytkowski et al. (2004); other CSP-ic results – *triangles*: Q_{el}; *stars*: Q_{ion}; *squares*: ΣQ_{exc}

The total cross sections reported in Rahman et al. (2012) were Q_T, Q_{el}, Q_{ion} and ΣQ_{exc}, with experimental measurements on Q_{ion} (with 15% uncertainty). For the weakly polar NF$_3$, the dipole rotational TCS are very small at energies above 10 eV. As shown in Figure 5.1 the TCS Q_T calculated within the CSP-ic formalism agree with the measured cross sections of Szmytkowski et al. (2004) across the energy range of their measurements.

While the TCS Q_T are in a general accord with measurements on NF$_3$, there is discrepancy between theories and experiments for the Q_{ion} for this target. As shown in Figure 5.2, the theoretical ionization TCS derived using CSP-ic, although closer to BEB results (Szmytkowski et al. 2004), do not agree with the measured data reported in Rahman et al. (2012). The earlier Q_{ion} measurements of Tarnovsky et al. (1994) and Haaland et al. (2001) are also much lower than the CSP-ic values, while the DM values of Deutsch et al. (1994) are closer to the experimental data in Rahman et al. (2012). Around 120 eV the Q_{ion} calculated in CSP-ic is 5.07 Å^2 while the experimental value reported by Rahman et al. (2012) is 3.6 Å². Most recently, Hamilton et al. (2017) employed ab initio R-matrix calculations augmented by other procedures, to calculate cross sections for electron collisions with NF_3, NF_2 and NF with potential applications to remote plasma sources. The theoretical Q_{ion} results of Hamilton et al. (2017), which include total including dissociative ionization cross sections on NF_3, are also larger than earlier measurements of Tarnovsky et al. (1994) and are closer to theoretical and experimental results of Rahman et al. (2012).

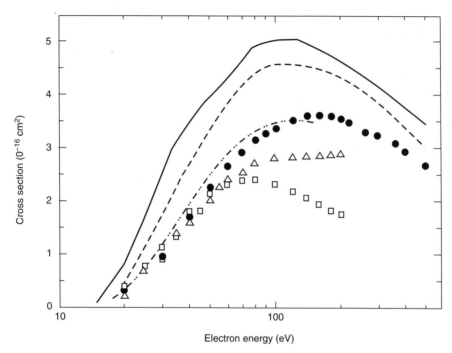

Figure 5.2 Ionization cross sections of NF_3; *squares*: Tarnovsky et al. (1994); *triangles*: Haaland et al. (2001); *filled circles*: measurements, Rahman et al. (2012); *solid line*: CSP-ic theory, Rahman et al. (2012); *dashed line*: BEB values, Szmytkowski et al. (2004), *dash-dotted line*: DM, Deutsch et al. (1994)

It is observed that for many of the fluorine containing compounds, the measured Q_{ion} are considerably lower than the theoretical results. This was confirmed in the unpublished CSP-ic calculations of Limbachiya (2016)[iii] on the radicals NF and NF_2. CSP-ic results (Limbachiya 2016)[iv] along with comparisons are depicted in Figure 5.3 for NF and in Figure 5.4 for NF_2. As can be seen here, the measured Q_{ion} in both NF and NF_2 are substantially lower (even beyond the error bars) than the CSP-ic and other theories.

---

iii, iv   Limbachiya, C. G. 2016. Private communication (unpublished).

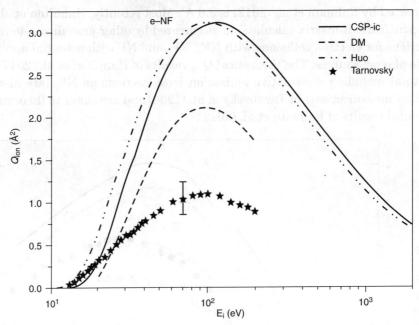

Figure 5.3   Electron scattering cross sections Q_{ion} with NF; *solid line*: CSP-ic, Limbachiya (2016)[v]:
dash: DM formula, Deutsch et al. (1994); *dash dot dot*: theory, Huo et al. (2002); *star*:
measured data of Tarnovsky et al. (1994)

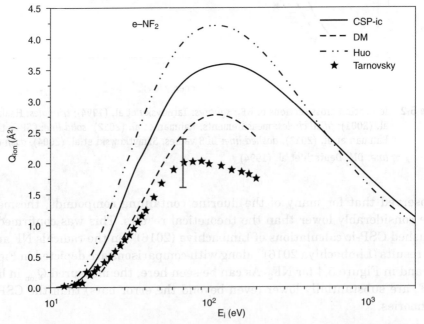

Figure 5.4   Electron scattering cross sections Q_{ion} with NF$_2$; *solid line*: CSP-ic, Limbachiya (2016)[vi];
dash: – DM formula, Deutsch et al. (1994); *dash dot dot*: theory, Huo et al. (2002); *star*:
measured data of Tarnovsky et al. (1994)

---

v, vi   See foot note above *iii, iv* (page 133)

The recent paper of Hamilton et al. (2017) reports an extensive set of cross sections for electron scattering processes on NF, NF_2 and NF_3 together with total and dissociative ionization cross sections for the radicals NF_x. Comprehensive calculations on electron–NF_3 scattering, ranging from 1 eV to 5 keV of energy were reported by Goswami et al. (2013). High energy calculations based on a modified Additivity Rule for Q_T were reported on NF_3, PF_3 and other polyatomics by Shi et al. (2010).

5.1.2 CF_4 and radicals CF_x (x = 1, 2 & 3)

Carbon tetrafluoride or CF_4, also called tetrafluoromethane, is a man-made gas with a diverse range of technological applications. CF_4 is the simplest stable fluorocarbon, with a rather large bond length $R_{C-F} = 1.31$ Å, an ionization energy I = 14.7 eV, and $\alpha_0 = 2.82$ Å^3 (CCCBDB database). It is widely used as a feed-gas in plasma assisted materials processing applications and therefore has been the subject of many electron collision studies. Several reviews of the status of the database for electron scattering from CF_4 have been compiled (Christophorou et al. 1996, Christophorou and Olthoff 1999), while Karwasz et al. (2014) presented a comprehensive review of experiments and binary-encounter models for electron impact ionization of fluoromethanes including CF_4. Total and ionization cross sections of a full set of plasma etching molecules CF_4, C_2F_4, C_2F_6, C_3F_8 and CF_3I and their radicals CF_x (x = 1–3) were calculated using the CSP-ic approximation by Antony et al. (2005). For targets like CF_4 the total complex potential was obtained from the single-centre charge density constructed at the molecular mass centre. The total (complete) cross sections Q_T of CF_4, calculated by Antony et al. (2005) are not displayed here, but they reproduced the experimental data above 100 eV while the theory appeared to overestimate Q_T at lower energies.

Figure 5.5 shows the status of ionization results in e–CF_4 scattering. The CSP-ic values (Antony et al. 2005) are in good agreement with the measurements of Nishimura et al. (1990), except at the peak where the experimental values are a little lower (but still within experimental uncertainty), while the measured data of Poll et al. (1992) are found to be higher than theoretical predictions and Nishimura et al. (1990). The BEB and the recommended Q_{ion} of Christophorou and Olthoff (1999) agree well with CSP-ic data. The peak Q_{ion} value of CF_4 in CSP-ic (Antony et al. 2005) is 5.7 Å^2 as against the more recent value 5.4 Å^2 given in the recommended data of Karwasz et al. (2014).

Since all of the electronically excited states of tetrahedral molecules CH_4 and CF_4 are repulsive in nature it has been argued that, for these targets, the total dissociation cross section denoted by Q_{Ndiss} equals the total electronic excitation cross section. It is in this context that the electronic-sum ΣQ_{exc} as estimated in the CSP-ic can provide new insights into excitation and dissociation of these molecules. In an earlier calculation, Joshipura et al. (2004) showed agreement between ΣQ_{exc} from CSP-ic with the neutral dissociation cross section as given by Christophorou and Olthoff (1999). In such cases, we have

$$Q_{inel} = Q_{ion} + Q_{Ndiss}$$

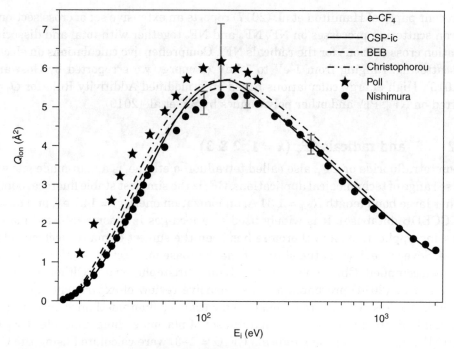

Figure 5.5 Total ionization cross sections of CF_4; *solid curve*: CSP-ic, Antony et al. (2005), *dash*: BEB (NIST database); *dash-dot*: Christophorou and Olthoff (1999); *star*: experimental data, Poll et al. (1992); *filled circle*: Nishimura et al. (1999)

Antony et al. (2005) also presented CSP-ic cross sections of radicals CF_x (x = 1, 2 and 3) vis-à-vis the parent stable molecule CF_4. The bond lengths in the three radicals CF_x are almost similar to those in CF_4 but in the radicals the thresholds of first ionization are lower.

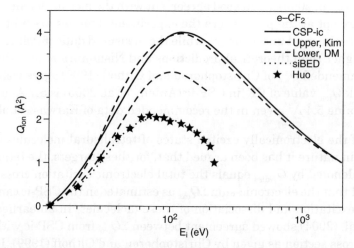

Figure 5.6 Total ionization cross sections of the CF_2 radical; *solid curve*: CSP-ic (Antony et al. 2005); *dashes*: BEB (NIST database); *dash-dots*: DM (Deutsch et al. 2000); *stars*: experimental data (Huo et al. 2002); *dots*: siBED model (Huo et al. 2002)

As a typical example of the discrepancies observed between electron impact ionization theory and experiment, we reproduce here, in figure 5.6, the Q_{ion} results (Antony et al. 2005) along with comparisons for CF_2. The CSP-ic theory in this case agrees with the DM values (Deutsch et al. 2000) and the BEB results (NIST database) but the experimental data of Huo et al. (2002) are far lower than the theoretical results (figure 5.6). Nevertheless the theoretical estimates in the simplified Binary Encounter Dipole (or siBED) model, given in Huo et al. (2002) reproduce their (Huo et al. 2002) experimental data. A similar discrepancy was also observed by Antony et al. (2005) for the other two radicals CF and CF_3 (not shown).

5.1.3   H$_2$CO and CHO

Formaldehyde H_2CO (or HCHO) is a toxic gas and a known health hazard. It exists in our upper atmosphere and was the first organic polyatomic molecule detected in the interstellar medium. H_2CO has been proposed as a key molecule in pre-biotic evolution because of the ease with which it may be formed under simulated conditions and its ability to condense with itself to form carbohydrates (Quayle and Ferenci 1978).

The basic properties of this molecule are I = 10.885 eV, D = 0.917 au and α_0 = 2.77 Å^3. Formaldehyde is hard to handle experimentally, and therefore much of the data on electron interactions with this molecule is theoretical. An extensive list of total cross sections was investigated in CSP-ic by Vinodkumar et al. (2006) and the problem was revisited by Vinodkumar et al. (2011). However, in these studies, the dipole rotational cross sections were not included and were later provided by Gangopadhyay (2008), wherein the grand total cross section was defined as $Q_{TOT} = Q_T + Q_{rot}$, with the rotational excitation cross section Q_{rot} calculated in the cut-off Born-dipole approximation (see Chapter 2, Section 2.3). Figure 5.7 shows Q_{TOT}, Q_{rot} and Q_{ion} for formaldehyde (Gangopadhyay 2008). Inclusion of Q_{rot} as mentioned, results in a very large TCS Q_{TOT} at lower energies. The CSP-ic Q_{ion} of this molecule (lower curves Figure 5.7) are in very good accord with the BEB values of Kim and co-workers (see NIST database).

Figure 5.8 shows the ionization results of H_2CO using the improved CSP-ic (ICSP-ic) calculations of Vinodkumar et al. (2011) and compares these with other data including the experiment of Vacher et al. (2009). Vacher et al. (2009) data suggest a Q_{ion} that peaks at a lower energy then the theoretical predictions. The earlier CSP-ic results of Vinodkumar et al. (2006) are also shown in this figure and agree well with ICSP-ic, albeit with a slightly lower peak cross section. The BEB derived Q_{ion} are in excellent agreement with ICSP-ic values at all energies.

In the case of the formyl radical (CHO), CSP-ic calculations on Q_T were carried out by Vinodkumar et al. (2006) and were augmented recently with the Born-dipole Q_{rot} (Joshi 2017)[vii].

---

vii   Joshi, Foram M. 2017. private communication (unpublished).

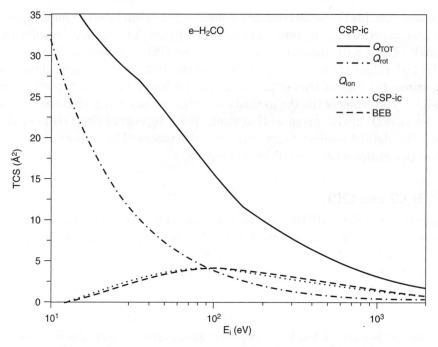

Figure 5.7   Theoretical cross sections Q_{TOT}, Q_{rot}, and Q_{ion} for H_2CO

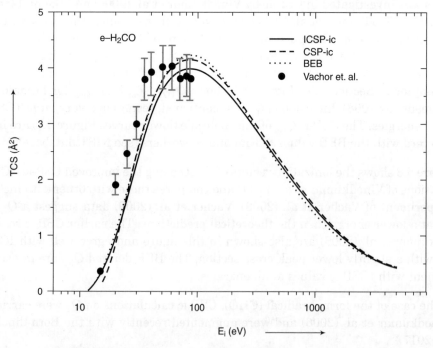

Figure 5.8   Total ionization cross sections of H_2CO; *continuous curve*: ICSP-ic (Vinodkumar et al. 2011); *dash-dots*: CSP-ic (Vinodkumar et al. 2006); *dashes*: BEB (NIST database); *bullets*: experimental data (Vacher et al. 2009)

Figure 5.9 displays various total cross sections for e–CHO scattering. The CSP-ic Q_{ion}, (not affected by long-range dipole interactions) as obtained by Vinodkumar et al. (2006) were in good agreement with the BEB results (NIST database). Recently Yadav et al. (2017) have derived electron scattering cross sections for the formyl radical at incident energies ranging from 0.01 to 5000 eV. They investigated a variety of processes and reported data on vertical electronic excitation energies, dissociative electron attachment (DEA) and total cross sections along with scattering rate coefficients. Yadav et al. also observed a Ramsauer–Townsend minimum at low energies.

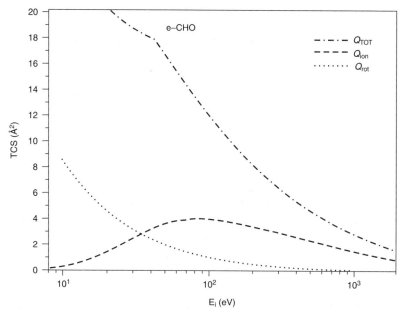

Figure 5.9   Total cross sections of the CHO radical; CSP-ic results (Vinodkumar et al. 2006) are augmented with the Born-dipole Q_{rot} (Joshi 2017)[viii] to obtain Q_{TOT}; Q_{ion} is that of Vinodkumar et al. (2006)

Vinodkumar et al. (2006) extended CSP-ic calculations to formic acid (HCOOH) but did not include the dipole rotation effects on Q_T. The Q_{ion} results for HCOOH (not shown here) were obtained later by Vinodkumar et al. (2011) with an approximately derived value $R_p = 0.50$ in CSP-ic, and were found to be in good agreement with the earlier experimental and theoretical data.

Finally, this sub-section covers electron scattering results for a molecule similar to formaldehyde, namely thioformaldehyde or H_2CS. Low energy calculations on H_2CS were performed by Limbachiya et al. (2015) using the R – matrix scattering code (Tennyson 2007, Tennyson et al. 2010) while the CSP-ic method was employed above 20 eV. The Q_{ion} calculated in CSP-ic by Limbachiya et al. (not shown here) are quite close to the corresponding BEB results. At their respective peak positions the CSP-ic Q_{ion} of H_2CO (Figure 5.8) and H_2CS (Limbachiya et al. 2015) are 4.42 Å^2 and 5.01 Å^2 respectively. There are no other theoretical or experimental data for H_2CS.

---

viii   See foot note above *vii* (page 137)

5.2   TWO-CENTRE SYSTEMS, SMALL HYDROCARBONS, etc.

Molecular targets such as the hydrocarbons C_2H_2 or C_2H_4 are two-centre systems in the sense that the two carbon atoms in such a molecule can be considered as two approximately independent scattering centres at high enough energy. Referring to Table 5.1 we find that the hydrocarbons C_2H_2, C_2H_4 and C_2H_6 are characterized by relatively larger C-C bond lengths compared to the C-H bond lengths. Therefore, a group additivity rule can be applied at high enough energies.

Table 5.1   Bond lengths in two-centre hydrocarbon targets (CCCBDB database)

Target	Bond length R_{C-H} (Å)	Bond length R_{C-C} (Å)
C_2H_2	1.063	1.203
C_2H_4	1.086	1.339
C_2H_6	1.091	1.536

Thus, we can start by constructing the electron charge-density of the functional (or constituent) group CH in C_2H_2, centred at its carbon atom and calculate various total cross sections in CSP-ic. This leads to the TCS of C_2H_2 being twice that of CH. A small modification is now introduced in the group-AR (Vinodkumar et al. 2006) by employing the actual molecular properties e.g. bond length of CH in C_2H_2 and the ionization threshold of the molecule, not the constituent. However, the AR still leads to an overestimation of the cross sections at intermediate and low energies and so a screening or overlap correction is required. This can be done in two approximate ways, as discussed in connection with two-centre targets such as HO_2 and H_2O_2 in Chapter 4. In the present case of C_2H_2, the static or geometrical correction or *sc1* (Chapter 4, Section 4.3) predicts a flat overestimation of about 8% in AR at all energies. The dynamic correction or *sc2* depends on incident energy and also on the magnitude of the TCS of the constituent itself. According to the correction sc2, the resulting overestimation in the peak Q_{ion} is about 8% in AR, while the sc2 predicts the AR for Q_T to be as high as 15-20 % at a low energy of about 20 eV. We can conclude from the bond lengths given in Table 5.1 that the group-AR gives more reliable results as we progress along the target sequence C_2H_2-C_2H_4-C_2H_6. In the results displayed here the screening corrections are not included but comparisons have been made with available data.

5.2.1 C_2H_2, C_2H_4, and C_2H_6

The linear molecule acetylene (ethyne, C_2H_2) is a simple unsaturated alkyne with a triple bond between the two carbon atoms. Acetylene has applications in various plasma processes. Acetylene along with di-acetylene has also been found to be of planetary and astrophysical interest. Measured electron scattering cross sections of the C_2H_2 molecule were compiled and reviewed very recently by Song et al. (2017). For C_2H_2, a hybrid theoretical study of electron scattering ranging from 1 to 5000 eV was carried out by Vinodkumar et al. (2012) who used the R-matrix method from 1 eV to the ionization threshold of the target, beyond which the SCOP formalism was employed. Vinodkumar et al. (2012) observed a consistent matching of the Q_T calculated in both the theories, at the ionization threshold.

Earlier, Vinodkumar et al. (2006) carried out intermediate and high energy CSP-ic calculations on electron scattering with C_2H_2 and other hydrocarbons. In Section 4.6 of Chapter 4 we included C_2H_2 as a member of the 14-electron molecules and the ionization cross sections were compared with relevant data (Pandya et al. 2012). Figure 5.10 depicts the TCS Q_T and Q_{ion} of acetylene along with other available data. Above 60 eV the CSP-ic results on Q_T (Vinodkumar et al. 2006) are slightly on the lower side compared to the measured data of Ariyasinghe (2002) and Sueoka and Mori (1989). Ionization cross sections in this figure are in a very good accord with the BEB data (NIST database) and also with the measurements of Hayashi (1990) and Gaudin and Hagemann (1967).

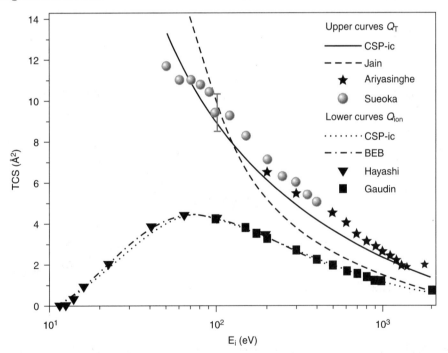

Figure 5.10 TCS of e–C_2H_2 scattering; *upper curves*: Q_T, continuous CSP-ic (Vinodkumar et al. 2006); *dashes*: theory, Jain and Baluja (1992); *stars*: experimental Q_T, Ariyasinghe (2002); *bullets*: Sueoka and Mori (1989); Q_{ion}: *lower dotted curve*: CSP-*ic* Vinodkumar et al. (2006); *dash dots*: BEB (NIST database); *inverted triangles*: Hayashi (1990); *boxes*: Gaudin and Hagemann (1967)

The recent review paper of Song et al. (2017) creates an opportunity to update comparisons of different data sets as in Table 5.2. The CSP-ic results (Vinodkumar et al. 2006) for all the three cross sections Q_T, Q_{el} and Q_{ion} are slightly lower than the recommended data of Song et al. (2017).

Table 5.2 TCS Q_T, Q_{el}, and Q_{ion} of C_2H_2 all in $Å^2$; (a) recommended data, Song et al. (2017) with uncertainties in the bracket each, (b) CSP-ic results (Vinodkumar et al. 2006)

Q_T		Q_{el}		Q_{ion}	
(a)	*(b)*	*(a)*	*(b)*	*(a)*	*(b)*
(11.5 ± 1.2)	9.8	(5.08 ± 1.33)	3.71	(5.32± 0. 48)	4.33

Vinodkumar et al. (2006) extended their CSP-ic calculations to many other hydrocarbons C_2H_4, C_2H_6, C_3H_4, C_3H_6, and C_3H_8 and the radicals $CH_x (x = 1–3)$. Consider the CSP-ic cross sections of e–C_2H_4 scattering, displayed in Figure 5.11, where their results are compared with other data. The CSP-ic Q_T of Vinodkumar et al. tend to agree with the experimental data of Sueoka and Mori (1989) above 100 eV but not at lower energies. The Q_{ion} derived from CSP-ic are seen to agree with experimental data (within stated uncertainties) and results from the BEB formalism.

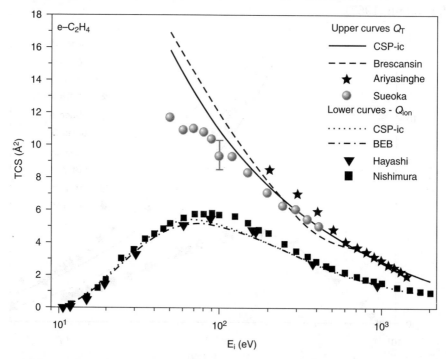

Figure 5.11 TCS for e–C_2H_4 scattering; *upper curves*: –Q_T, continuous CSP-ic (Vinodkumar et al. 2006); *dashes*: theoretical results (Brescansin et al. 2004); *stars*: experimental Q_T, (Ariyasinghe and Powers 2002); *bullets*: Sueoka and Mori (1989); *lower plots*: Q_{ion}; *dotted curve*: CSP-ic, (Vinodkumar et al. 2006); *dash-dots*: BEB (NIST database); *inverted triangles*: Hayashi (1990); *squares*: Nishimura and Tawara (1994)

Figure 5.12 displays the TCS results for ethane C_2H_6, a very common organic compound that has been widely studied in electron scattering both in theory and experiment. Total (complete) cross sections Q_T derived using CSP-ic (Vinodkumar et al. 2006) are slightly higher than the measured data of Sueoka and Mori (1986) which is available up to 400 eV. Total elastic cross sections Q_{el} are not shown in this figure but from the CSP-ic results (Vinodkumar et al. 2006) one finds that at 1000 eV, $Q_{el} \approx Q_T - Q_{ion} = 2.34$ $\mathring{A}^2$, somewhat larger than $Q_{ion} = 1.47$ $\mathring{A}^2$ at that energy.

Note that at higher energies the measured Q_T data of Ariyasinghe and Powers (2002) are progressively lower than the theoretical values (Figure 5.12). This behaviour is difficult to explain especially since the experimental data of Ariyasinghe and Powers (2002) tend to lie close to the ionization cross sections (Figure 5.12) suggesting a very low total elastic cross sections Q_{el} which is physically unfeasible. Sum-check issues such as this can be identified by presenting an all-TCS plot of electron scattering from a particular target, and that is the practice we follow for many of the targets discussed in the present monograph, where we use such comparisons to identify self-consistent sets of data.

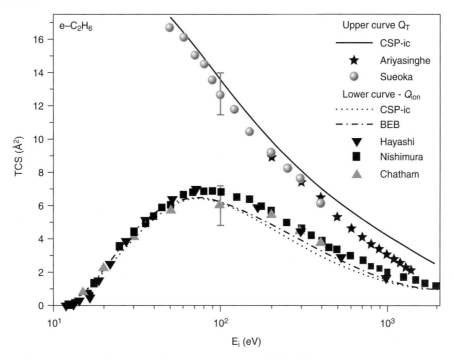

Figure 5.12 TCS of e–C_2H_6 scattering; *upper curve*: Q_T, CSP-ic (Vinodkumar et al. 2006); *bullets*: Sueoka and Mori (1986): *stars*: Ariyasinghe and Powers (2002); *lower curves*: Q_{ion}, *dotted curve*: CSP-*ic* Vinodkumar et al. (2006); *dash-dots*: BEB (NIST database); *inverted triangles*: Hayashi (1990); *squares*: Nishimura and Tawara (1994); *triangles*: Chatham et al. (1984)

Electron impact ionization cross sections on C_2H_6 are displayed in the lower part in Figure 5.12. There is a good accord between the two theories CSP-ic and BEB while the experimental data are close to these two theories up to the Q_{ion} peak, above which the measured data (Figure 5.12) tend to be somewhat higher.

An interesting empirical relationship, based on the measured data, between the TCS Q_T and Q_{ion} for a variety of simple hydrocarbons and perfluorocarbons was proposed by Kwitnewski et al. (2003), who showed that at high energies the ratio Q_{ion}/Q_T tends to a particular value (around 1/2 or 1/3) for a given molecule. For C_2H_6 and other simple hydrocarbons, the ratio assumes a high-energy value close to 0.5, indicating equal contributions of Q_{el} and Q_{ion} in total electron scattering.

5.2.2 C_2F_2, C_2F_4, and C_2F_6

We begin a study on two-centre fluorocarbons by displaying in Table 5.3 the C-F and the C-C bond lengths of C_2F_2, C_2F_4, and C_2F_6. A recent theoretical study on e–C_2F_2 scattering may be found in Gupta et al. (2017). The next molecule in the series C_2F_4 is somewhat unusual as it exhibits nearly equal bond lengths R_{C-F} and R_{C-C} (Table 5.3). One would expect a large effect of overlap or screening in this case. However, the group-AR CSP-ic calculations (Antony et al. 2005) yielded reasonably good results for ionization cross sections of C_2F_4 without any screening corrections.

Table 5.3 Bond lengths in the two-centre fluorocarbon targets (CCCBDB database)

Target	Bond length R_{C-F} (Å)	Bond length R_{C-C} (Å)
C_2F_2	1.187	1.283
C_2F_4	1.319	1.311
C_2F_6	1.32	1.56

As shown in Figure 5.13, the CSP-ic results for Q_{ion} of C_2F_4 are close to the experimental data (Bart et al. 2001) up to the peak position. The peak Q_{ion} obtained by Antony et al. (2005) was 5.90 Å^2, which is comparable with the corresponding result 6.16 Å^2 given in the recommended data of Rozum et al. (2006). Moreover, the results of Antony et al. (2005) are found to be in accord with the recent BEB (RHF) results of Gupta et al. (2017).

Hexafluoroethane C_2F_6, the fluoro-counterpart of the hydrocarbon C_2H_6, is an extremely stable greenhouse gas and is widely used in the plasma industry and thus the subject of many experimental and theoretical studies. Recommended data for electron scattering from it was first presented by Christophorou et al. (1996) and Christophorou and Olthoff (1998). Figure 5.14 reveals good agreement between the CSP-ic Q_{ion} results of Antony et al. (2005) and the recommended data of Christophorou and Olthoff (1998), the BEB values (NIST database) as well as the experimental data of Nishimura et al. (2003). However, the experimental data of Poll et al. (1992) lies below the other results above 80 eV. The fluorocarbon C_2F_6 is also one of the targets included in the recent theoretical investigations of Gupta et al. (2017) in their analysis of different versions of the BEB theory. The CSP-ic results (Antony et al. 2005) are found to agree well with the data presented in Gupta et al. (2017).

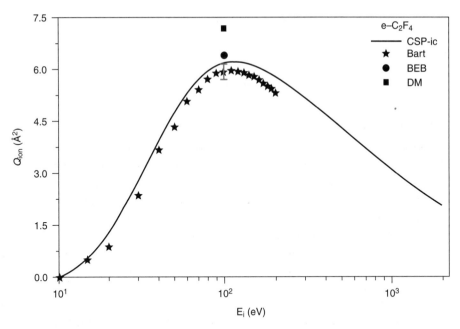

Figure 5.13 Ionization cross sections of C_2F_4; *solid curve*: CSP-ic (Antony et al. 2005); *single bullet*: BEB (NIST database); *single square*: DM from Bart et al. (2001); *stars*: experimental data, Bart et al. (2001)

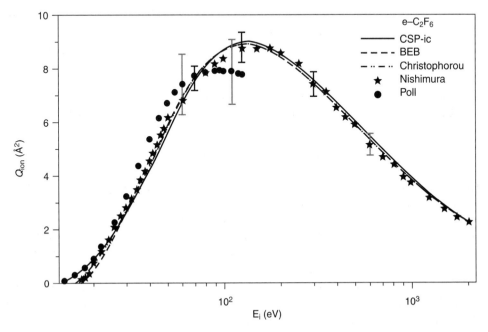

Figure 5.14 Ionization TCS of e–C_2F_6; *continuous curve*: CSP-ic (Antony et al. 2005); *dashes*: BEB (NIST database); *stars*: Experimental data – Nishimura et al. (2003); *bullets*: Poll et al. (1992); *dash-dots*, Christophorou and Olthoff (1998)

Therefore despite the concern that the AR may breakdown for such targets it is seen that for stable fluorine compounds there is a good general agreement between theory and experiment.

Now let us explore a few other two-centre targets.

5.2.3 CH_3OH, CH_3I, and CF_3I

One of the most important compounds formed by the methyl group is methyl alcohol CH_3OH. The molecular structure suggests that the application of group-AR in this target is valid. Vinodkumar et al. (2011) obtained the interaction potential V_{abs} with a parametric expression of the threshold energy Δ, and hence calculated the Q_{ion} of CH_3OH, by employing the ICSP-ic method, with a ratio parameter $R_p = 0.74$. The results of Vinodkumar et al. are not reproduced here but their ionization cross sections of methanol molecule are found to be in a good agreement with available experimental data and BEB results. Vinodkumar et al. also reported theoretical results for C_2H_5OH and 1-C_3H_7OH molecules. The dipole moment of methanol molecule is 0.66 D and the two other alcohols are also dipolar. Therefore, it would be desirable in these calculations to obtain the grand total cross sections Q_{TOT} and rotationally inclusive TCS Q_{el}. A compilation of various electron scattering cross sections with these and other molecules of interest in the field of gaseous electronics, has been performed by Govinda Raju (2011).

Briefly on the subject of an organo-sulphur compound CH_3SH, called methanethiol; CSP-ic was employed by Limbachiya et al. (2014) to obtain total cross sections of e–CH_3SH scattering from threshold to 5 keV but it was noted that there is a lack of other data with which to compare.

With the expectation that a group-AR should yield reliable results for methyl halides, the tabulated results of Naghma and Antony (2013) show that at 100 eV the respective Q_{ion} decrease in magnitude according to

$$CH_3I > CH_3Br > CH_3Cl > CH_3F$$

However these theoretical results differ somewhat from the measured data of Rejoub et al. (2002), in terms of both peak position and magnitude.

It is of interest here to include trifluoroiodomethane CF_3I, a tetrahedral polar molecule with bond lengths $R_{C-F} = 1.329$ Å and $R_{C-I} = 2.144$ Å. An important potential candidate to replace fluorocarbons in plasma etching (with lower global warming potential) electron interactions with this target were reviewed by Christophorou and Olthoff (2000) and updated by Rozum et al. (2006). CSP-ic calculations on CF_3I were performed by Antony et al. (2005) and by Vinodkumar et al. (2006) both of which agreed in their calculated Q_T and were in good agreement with experimental data above 70 eV. In Figure 5.15 we focus on two different evaluations of Q_{ion}. Theoretical DM values from Onthong et al. (2002) are seen to be consistent with CSP-ic derived values by Antony et al. (2005) up to the peak position, above which the DM data are found to be lower than CSP-ic data and this is observed in many other cases. Measured data of Jiao et al. (2001) available up to the Q_{ion} peak are close to both of the theories. More recently experimental elastic integral cross sections (Q_{el} in our notation) for the CF_3I molecule were reported by Hargreaves et al. (2011).

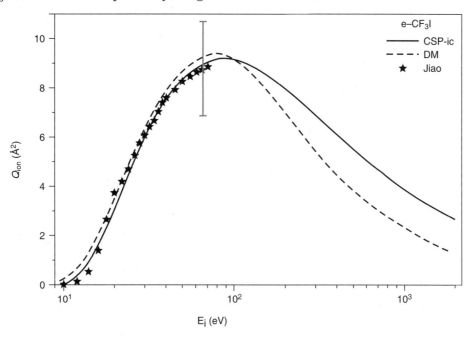

Figure 5.15 Total ionization cross sections of e–CF_3I scattering; *solid curve*: CSP-ic (Antony et al. 2005); *dashes*: DM (Onthong et al. 2002); *stars*: experimental (Jiao et al. 2001)

5.3 LARGER TETRAHEDRAL AND OTHER MOLECULES

5.3.1 SiH_4 and radicals SiH_x (x = 1,2, and 3)

The tetrahedral nonpolar molecule silane SiH_4 (bond length R_{Si-H} = 1.48 Å) is often compared to methane in terms of its electron scattering properties. SiH_4 has numerous applications in natural and technological plasmas. In such environments, where plasma assisted processes are vital, electron induced reactions play a major role in its chemistry.

Due to its symmetric structure a single-centre charge density and hence total complex potential can be constructed to derive various cross sections. Figure 5.16 displays total (complete) cross sections Q_T of SiH_4. Vinodkumar et al. (2008) computed CSP-ic for silane and other silicon compounds from threshold to 5 keV. Good agreement is found between the CSP-ic and the two experimental measurements Zecca et al.

(1992), and Szmytkowski et al. (1995). The theoretical results of Jain and Baluja (1992) and Jiang et al. (1995) are however seen to be higher and lower respectively.

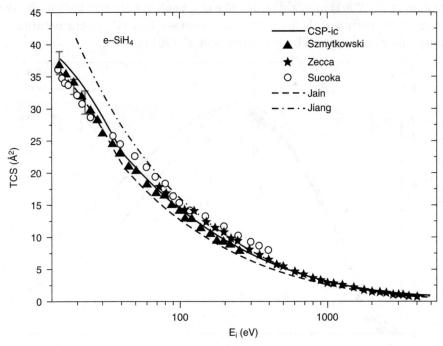

Figure 5.16 Total (complete) cross sections Q_T of silane; *line*: theory; continuous curve CSP-ic of Vinodkumar et al. (2008); *dash*: Jain and Baluja (1992); *dashed dot line*: Jiang et al. (1995); *experimental stars*: Zecca et al. (1992); *triangles*: Szmytkowski et al., (1995); *open circles*: Sueoka et al. (1994)

Q_{ion} results for silane are shown in Figure 5.17. The CSP-ic Q_{ion} results are seen to agree well with the BEB data of Ali et al. (1997) and with the measured data of Basner et al. (1997) and Chatham et al. (1984). However the Q_{ion} results of Krishnakumar and Srivastava (1995) and Malcolm et al. (2000) show considerable differences. Zecca et al. (2001) have made an important observation that in the electron impact ionization of SiH_4 no stable parent ions (SiH_4^+) were observed in earlier experiments and since fragments are produced with the high kinetic energies, this may lead to some systematic effects in experimental data (see Chapters 2, 3).

Vinodkumar et al. (2008) represented the molecular cross section Q_T (in a_0^2) as a function of incident energy E_i (in eV) and average molecular polarizability α_0 (in a_0^3) by a simple formula in terms of fitting parameters A and B, as follows.

$$Q_T = A\left(\frac{\alpha_0}{E_i}\right)^B \tag{5.2}$$

For SiH_4, Vinodkumar et al. found the parameters to be $A = 117.39$ and $B = 0.72$.

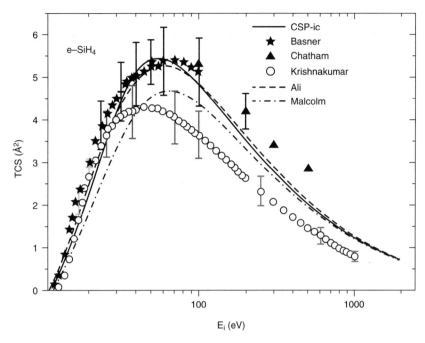

Figure 5.17 Ionization cross sections of SiH$_4$; *continuous curve*: theory, CSP-ic (Vinodkumar et al. 2008), *dashes*: BEB, (Ali et al. 1997); *chain curve*: Malcolm et al. (2000); *stars*: experimental, Basner et al. (1997); *triangles*: Chatham et al. (1984); *open circles*: Krishnakumar and Srivastava (1995)

Table 5.4 Various theoretical TCS (in Å^2) for SiH$_4$ (a) Gangopadhyay (2008) (b) Vinodkumar et al. (2008) (c) Verma et al. (2017) (d) BEB (NIST database)

Q_T			Q_{el}	Q_{ion}		ΣQ_{exc}
(a)	*(b)*	*(c)*	*(a)*	*(a)*	*(d)*	*(a)*
17.75	14.58	13.85	11.40	4.93	5.24	1.42

Table 5.4 shows the TCS of e–SiH$_4$ scattering at 100 eV for various theoretical calculations and there is general agreement amongst the various datasets. A very recent work on electron–silane scattering is from Verma et al. (2017) who have employed the R-matrix theory in the quantemol-N package for low energy calculations and CSP-ic (with V$_{abs}$ in fixed-Δ model) for intermediate to high energies.

Consider next the electron impact ionization of the radicals SiH$_x$ (x = 1, 2 and 3) which are derived from the parent molecule. We have selected here sample results on silicon monohydride SiH, as in Figure 5.18. In this figure, the peak position in the experimental data of Tarnovsky et al. (1997) appears at 70 eV and is clearly different from the other results, which are all consistent with a rather low ionization threshold of SiH i.e. 7.89 eV.

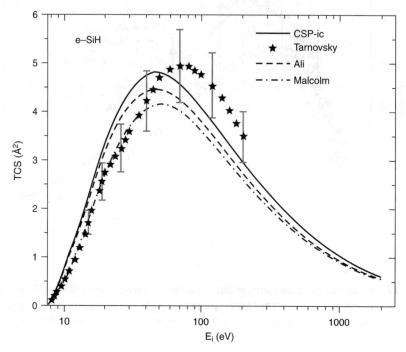

Figure 5.18 Q_{ion} of SiH; *continuous curve*: CSP-ic, Vinodkumar et al. (2008); *dashes*: BEB, (Ali et al. (1997); *dash-dots*: Malcolm and Yeanger (2000); *stars:* experimental data, Tarnovsky et al. (1997)

5.3.2 SiF$_4$, GeH$_4$, and radicals

Turning to electron scattering from silicon tetrafluoride SiF$_4$ (bond length $R_{Si-F} = 1.554$ Å), we note that Joshipura et al. (2004) had published the CSP-ic computations of this target's major TCS.

Figure 5.19 shows the Q_T and Q_{ion} of SiF$_4$ (Joshipura et al. 2004). No comparable data seems to be available for Q_T while the CSP-ic theoretical ionization cross sections are in agreement with the measured data of Basner et al. (2001). The Q_{ion} of SiF$_4$ exhibits a broad peak around 100 eV, similar to the BEB results (NIST database), not shown in this figure. The lowest curve represents ΣQ_{exc} which, since all electronic states of SiF$_4$ are dissociative in nature, is equal to the total neutral dissociation cross section at a particular energy.

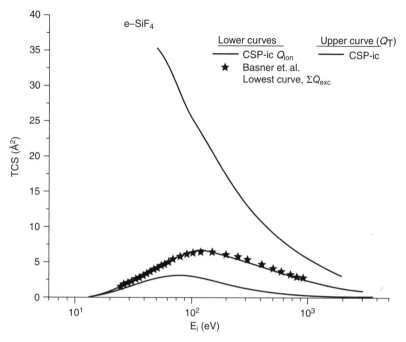

Figure 5.19 Set of cross sections for e–SiF$_4$ scattering; *upper curve*; Q_T; *lower curve*: Q_{ion} (Joshipura et al. 2004); *stars*: Q_{ion} experimental, Basner et al. (2001); *lowest curve:* ΣQ_{exc}

On the subject of the heavier and larger tetrahedral target germane GeH$_4$ (R$_{Ge-H}$= 1.53 Å); note that the bond lengths and polarizabilities of silane and germane molecules are nearly the same (CCCBDB database). Silane–germane gas mixtures are often of relevance in plasma studies. Total (elastic plus inelastic) cross sections, i.e., Q_T were evaluated in the SCOP formalism for GeH$_4$ (along with CF$_4$) by Baluja et al. (1992). Absolute measurements for total cross sections (Q$_T$) of GeH$_4$ in the range 0.75 – 250 eV were carried out by Mozejko et al. (1996). CSP-ic calculations for total and ionization cross sections of germanium hydrides (including GeH$_4$) were performed by Vinodkumar et al. (2008). For germane, we find that at 100 eV the calculated Q_T value 14.13 Å^2 is quite close to the measured value 14.9 Å^2 (Mozejko et al. 1996).

Various results on e–GeH$_4$ ionization are given in Figure 5.20 reproduced from Vinodkumar et al. (2008). The CSP-ic results are in good agreement with the BEB values of Ali et al. (1997), while the DM formula results (Probst et al. 2001) rise to a peak at lower energy and tend to fall off more rapidly at higher energies. The single-point experimental data of Perrin and Aarts (1983) has a large error bar, and is significantly different from the three theories (figure 5.20) but the single data point is supported by the Q_{ion} estimates of Szmytkowski and Denga (2001) based on a regression formula.

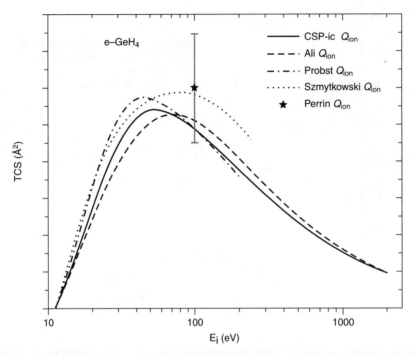

Figure 5.20 Ionization in e–GeH$_4$ scattering; theory, *full line curve*: CSP-ic (Vinodkumar et al. 2008); *dashes*: BEB (Ali et al. 1997); *dash-dots*: DM (Probst et al. 2001); *dots*: estimated Q_{ion} (Szmytkowski and Denga 2001); *star*: experimental, Perrin and Aarts (1983)

Table 5.5, contains the various theoretical TCS values for silane and germane at 100 eV, and these are found to be similar for the two molecules.

Table 5.5 Various theoretical TCS (in Å^2) for silane and germane at 100 eV, compiled from Vinodkumar et al. (2008)

Q_T		Q_{el}		Q_{ion}		ΣQ_{exc}	
silane	*germane*	*silane*	*germane*	*silane*	*germane*	*silane*	*germane*
14.58	14.13	8.53	7.86	4.85	4.88	1.20	1.39

Now, as an example of electron scattering from germanium hydride radicals, consider the plots for GeH displayed in Figure 5.21, showing the ionization results from Vinodkumar et al. (2008) together with the BEB calculations of Ali et al. (1997) and the DM method of Probst et al. (2001). Interestingly there are major differences in both the shape and peak values amongst the three theories.

Regarding GeF$_4$, a large tetrahedral molecule with R$_{Ge-F}$ = 1.67 Å having I = 15.5 eV (CCCBDB database), the total (Complete) cross sections Q_T for e–GeF$_4$ scattering

(0.5 to 250 eV) were measured by Szmytkowski et al. (1998), who found that at 100 eV, $Q_T = 23.4$ Å^2 for GeF_4 as against 14.9 Å^2 for GeH_4.

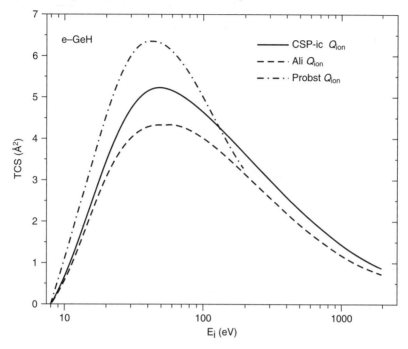

Figure 5.21 Total ionization cross sections of GeH; CSP-ic Vinodkumar et al. (2008); *dash curve*: BEB (Ali et al. 1997); *dash-dots*: DM (Probst et al. 2001)

5.3.3 CCl_4, $SiCl_4$, and radicals $SiCl_x$ (x = 1,2, and 3)

Carbon tetrachloride CCl_4 is a well-known compound that was used as a refrigerant for several decades until its global warming potential led to it being phased out. The basic properties of this molecule are $R_{C-Cl} = 1.767$ Å, ionization threshold I = 11.47 eV and a high polarizability $\alpha_0 = 10.002$ Å^3 (CCCBDB database). Figure 5.22 shows data on this molecule reproduced from Joshipura et al. (2004). In CSP-ic (with variable-Δ model in V_{abs}) the cross sections Q_T are consistent with the measured data of Zecca et al. (1991,1992) up to about 100 eV, below which the theory tends to be higher than experiment. Surprisingly the AR results of Jiang et al. (1995) are lower than both the CSP-ic and the measured data across the entire energy range. The CSP-ic ionization results are also in agreement with the measurements of Hudson et al. (2001) while for the electronic-sum ΣQ_{exc} no other data are available. Limão–Vieira et al. (2011) measured the elastic differential cross sections of this target in the energy interval 1.5 – 100 eV, and also derived the integral (total) elastic cross sections (i.e. Q_{el} in our notation). At 100 eV the Q_{el} of Limão–Vieira et al. (2011) is close to the Q_{ion} derived earlier by CSP-ic (Joshipura et al. 2004).We note for a comparison purpose that the magnitude of the peak of Q_{ion} for the CF_4 molecule in CSP-ic (Antony et al. 2005) is 5.7 Å^2, while that of CCl_4 (Joshipura et al. 2004) is much larger at 14.9 Å^2.

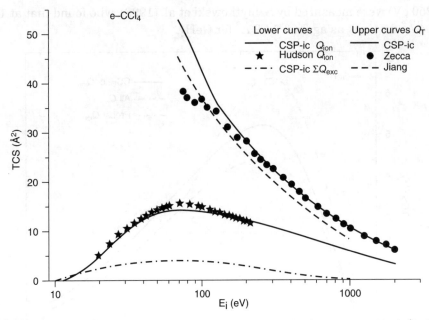

Figure 5.22   Cross sections for e–CCl$_4$ scattering; Q_T–*uppermost continuous curve:* CSP-ic (Joshipura et al. 2004); *dashes:* AR Jiang et al. (1995); *bullets:* experimental data Zecca et al. (1991,1992); *lower curves:* Q_{ion} – *continuous curve:* CSP-ic (Joshipura et al. 2004); *stars:* experimental data, Hudson et al. (2001); *lowest curve:* ΣQ_{exc}

A still larger tetrahedral target is SiCl$_4$ with a bond length of R$_{Si-Cl}$ is 2.019 Å, along with the ionization threshold I = 11.79 eV and polarizability α_0 = 11.27 Å^3 (CCCBDB database). CSP-ic studies on electron scattering with SiCl$_4$ and the radicals SiCl$_x$ (x = 1, 2 and 3) were conducted by Kothari et al. (2011). The radicals SiCl$_x$ are produced by the collisional break-up of the parent molecule SiCl$_4$. The SiCl$_3$ radical is a major chloro-silicon species involved in the chemical vapour deposition of silicon films from SiCl$_4$/Ar microwave plasmas. CSP-ic calculations (Kothari et al. 2011) indicate that in SiCl$_4$ the inelastic peak occurs at E$_p$ = 55 eV while the ionization peak occurs at 80 eV. Even at 100 eV the TCS of this molecule are quite high, namely

$$Q_T = 43.12 \text{ Å}^2, \; Q_{el} = 22.45 \text{ Å}^2, \text{ and } Q_{ion} = 17.58 \text{ Å}^2$$

Graphical plots of various TCS for silicon tetrachloride molecule (Kothari et al. 2011) are given in Figure 5.23. The CSP-ic Q_T are seen to be in agreement with the measurements of Mozejko et al. (1999) up to about 80 eV, below which the theory appears to overestimate the cross section. Mozejko et al. (1999) measured total (complete) cross sections of SiCl$_4$ over two different incident energy ranges of 0.3–250 eV and 75–4000 eV. There is a difference in the trend of these two data-sets in the overlapping region around 100 eV. Elastic cross sections Q_{el} of Kothari et al. are compared here with the results of Mozejko et al. (2002) which are derived from an Independent-Atom-model and are therefore higher than Kothari et al. data (Figure 5.23).

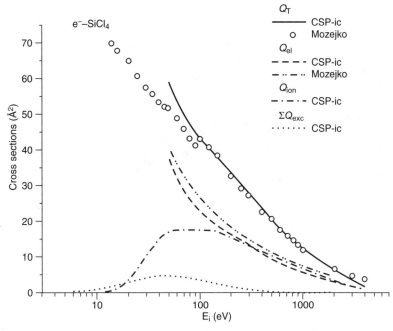

Figure 5.23   All TCS of $SiCl_4$; Q_T – *uppermost continuous curve*: CSP-ic (Kothari et al. 2011); *open circles*: experimental Mozejko et al. (1999); Q_{el} – *dashes*: CSP-ic (Kothari et al. 2011); *dash-double-dots*: Mozejko et al. (2002); *lower curves*: *dash dots* and *dots*: Q_{ion} and ΣQ_{exc} in CSP-ic

The ionization cross sections of $SiCl_4$ are depicted separately in Figure 5.24 adapted from Kothari et al. (2011). The CSP-ic Q_{ion} are in agreement, within the experimental uncertainties, with experimental data of Basner et al. (2005), while the measured partial ionization cross sections (PICS) of King and Price (2011) are lower throughout the energy range of their measurements. There are two DM calculations, one by Basner et al. (2005) and the other by Deutsch et al. (1997) and both these are at variance with CSP-ic and other data included in this figure. Notably, the partial ionization cross sections PICS of King and Price (2011) on $SiCl_4$ are close to the experimental Q_{ion} data on CCl_4 molecule as given by (Hudson et al. 2001).

Consider now the electron ionization of the radical $SiCl_3$ displayed in Figure 5.25 reproduced from Kothari et al. (2011). The CSP-ic results are in general on the higher side but near the peak region are close to the error bar in the Q_{ion} measurements of Gutkin et al. (2009). There are two DM calculations, those of Gutkin et al. (2009) are closer to CSP-ic up to the peak, beyond which the DM values of Gutkin et al. rapidly fall off with rising values of energy. The other DM results, namely those of Deutsch et al. (1997), give a much lower cross section at the peak (Figure 5.25).

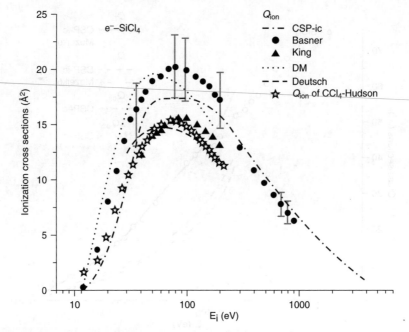

Figure 5.24 Electron impact ionization cross sections of SiCl$_4$; theory, *dash-dots* CSP-ic (Kothari et al. 2011), *dots* DM Basner et al. (2005), *dashes* DM Deutsch et al. (1997); experiment – *bullet*, Basner et al. (2005), *triangle*, PICS of King and Price (2011); *star*, measured data on CCl$_4$ molecule (Hudson et al. 2001)

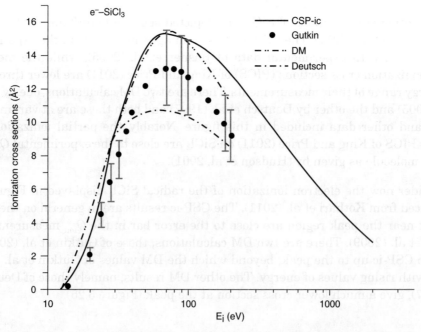

Figure 5.25 Ionization cross sections of SiCl$_3$ radical; *uppermost continuous curve*: theory-CSP-ic, Kothari et al. (2011); *dash-dots*: DM Deutsch et al. (1997); *dash-double-dots*: DM Gutkin et al. (2009); *bullets*: experimental data, Gutkin et al. (2009)

5.3.4 Si_2H_6, Ge_2H_6, and $Si(CH_3)_4$

Disilane Si_2H_6 although similar in structure to ethane C_2H_6, is much more reactive. In disilane the bond lengths are $R_{Si-Si} = 2.32$ Å, and $R_{Si-H} = 1.47$ Å, providing justification for treating this target as a two-centre scattering system. Experimental results on e–Si_2H_6 were reported by Chatham et al. (1984) and also by Krishnakumar and Srivastava (1995), while theoretical BEB results were reported by Ali et al. (1997) and DM results were given by Deutsch et al. (1998). CSP-ic calculations on disilane were carried out by Joshipura et al. (2007), and by Vinodkumar et al. (2008) with quite similar results.

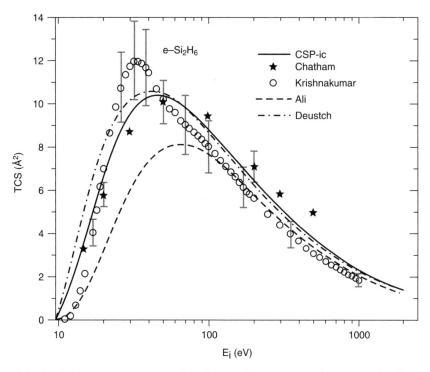

Figure 5.26(a)   Ionization cross sections of Si_2H_6; *continuous curve*: theory, CSP-ic (Vinodkumar et al. 2008), *dashes*: BEB Ali et al. (1997); *dash-dots*: DM Deutsch et al. (1998); measurements, *stars*: Chatham et al. (1984); *circles*: Krishnakumar and Srivastava (1995)

Figure 5.26(a) shows Q_{ion} results of Si_2H_6. The CSP-ic results are close to the measurements of Chatham et al. (1984) except at high energies, while the measured data of Krishnakumar and Srivastava (1995) are in general lower, except at the peak region where they are rather high. The DM values (Deutsch et al. 1998) are close to CSP-ic figure 5.26(a), but the BEB values (Ali et al. 1997) are distinctly lower below 100 eV. Just for comparison, at 60 eV, the CSP-ic Q_{ion} of SiH_4 is 5.44 Å^2 while that of Si_2H_6 is 10.32 Å^2.

Total (complete) cross sections for the disilane molecule were presented by Vinodkumar et al. (2008), and though not shown here, were found to be consistent with the corresponding experimental data, except below 30 eV. Moreover, Vinodkumar et al. represented Q_T (in a_0^2) as a function of E_i (in eV) and α_0 (in a_0^3), parametrically as in equation 5.2, and found that the parameters were A = 159.92 and B = 0.71 in this case.

Gases like germane, GeH_4 and digermane Ge_2H_6 are used as feed gases in plasma deposition and doping processes in the semiconductor industries. The structure properties of digermane Ge_2H_6 are, R_{Ge-Ge} = 2.445 Å, while R_{Ge-H} = 1.538 Å (Carrier et al. 2006). Despite their importance in plasma technologies there appear to be very few studies of electron scattering from such targets. CSP-ic calculations on this target were performed by Vinodkumar et al. (2008) and a BEB computation was reported by Ali et al. (1997); these are compared in figure 5.26(b). The magnitudes agree but the peak position of the cross section is different in the two models.

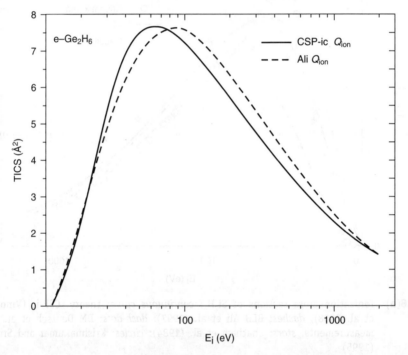

Figure 5.26(b)   Total ionization cross sections TICS of Ge_2H_6; *solid line*: CSP-ic, Vinodkumar et al. (2008); *dash*: BEB Q_{ion}, Ali et al. (1997)

Next, consider the organo-silicon compound called tetramethylsilane, $Si(CH_3)_4$, having bond lengths R_{C-H} = 1.115 Å, and R_{Si-C} = 1.875 Å. Electron impact ionization of this tetrahedral target (along with other silicon compounds) was investigated by Joshipura et al. (2007). As shown in Figure 5.27, the CSP-ic results are in reasonably good agreement with the measured data of Basner et al. (1996) up to their available results peaking at 100 eV. The BEB results of Ali et al. (1997) agree with CSP-ic in shape but are lower in magnitude around the peak region. The summed excited state cross section is a significant fraction of Q_{ion} below 100 eV.

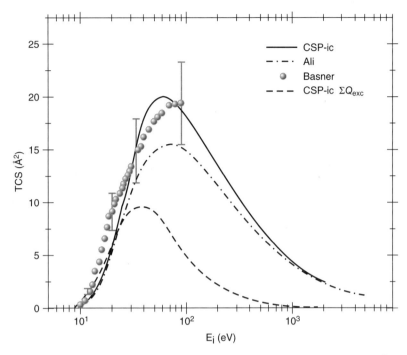

Figure 5.27 Electron collisions with $Si(CH_3)_4$; Q_{ion} theory, *continuous curve*: CSP-ic (Joshipura et al. 2007); *dash-dots*: BEB (Ali et al. 1997); *bullets*: experimental data (Basner et al. 1996); *dotted curve*: CSP-ic ΣQ_{exc}

5.4 HEAVIER POLYATOMICS

5.4.1 SO_2XY (X, Y = F, Cl)

Electron collisions with the two polar molecules SO and SO_2 were discussed in Chapter 4. Here we review data for electron scattering from exotic tetrahedral polar molecules SO_2XY (X, Y = F, Cl). Low to intermediate energy measurements of TCS Q_{TOT} on SO_2Cl_2 and SO_2F_2 were reported by Szmytkowski et al. (2006), while similar measurements on SO_2ClF were reported in Szmytkowski et al. (2005).

In SO_2Cl_2 molecule, the bond lengths are $R_{S-O} = 1.418$ Å and $R_{S-Cl} = 2.012$ Å. It is fairly ionizable and polarizable. The CSP-ic calculations of Joshipura and Gangopadhyay (2008) provided TCS Q_T, Q_{el} and Q_{ion}, while the grand total cross section is given by $Q_{TOT} = Q_T + Q_{rot}$, where Q_{rot} is the Born-dipole TCS; these are shown in Figure 5.28. The low energy theoretical Q_{TOT} in this figure (topmost curve) is quite high in view of the dipole rotational contribution. However, this is not reflected in the measured Q_{TOT} of Szmytkowski et al. (2006), which are closer to CSP-ic theory in the intermediate energy range. At energies exceeding 70 eV, the experimental Q_{TOT} data lie above the

CSP-ic values. The theoretical Q_{el} (middle curve) is plotted without including Q_{rot} while the Q_{el} of Szmytkowski et al. (2006), estimated using the Independent Atom Model are higher, as expected. Q_{ion} determined in CSP-ic agree with BEB data given by Szmytkowski et al. (2006).

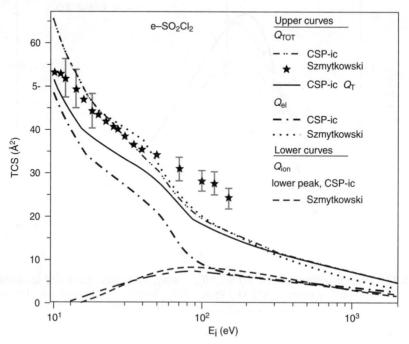

Figure 5.28 Electron scattering with SO_2Cl_2; Q_{TOT} – *uppermost dash-double dot curve*: CSP-ic (Joshipura and Gangopadhyay 2008); *stars*: experimental data (Szmytkowski et al. 2006), Q_T – *solid curve*: CSP-ic; Q_{el} – *dash-dot curve*: CSP-ic; *dotted curve*: IAM (Szmytkowski et al. 2006); *lower curves*: Q_{ion}; *lower peak*: CSP-ic, *dash*: BEB (Szmytkowski et al. 2006)

Figure 5.29 displays a data compilation for SO_2ClF. CSP-ic results (Joshipura and Gangopadhyay 2008) may be compared with measured Q_{TOT} of Szmytkowski et al. (2005) but there is hardly any agreement. The Q_{el} of Szmytkowski et al. (2005) estimated using IAM are higher, especially at lower energies. The lowest two curves in this figure represent Q_{ion}, wherein the BEB data (Szmytkowski et al. 2005) tend to be higher than the CSP-ic values beyond the ionization peak.

The last member in this sequence is SO_2F_2 for which there appear to be hardly any experimental data. In Figure 5.30, the CSP-ic cross sections Q_{TOT} (inclusive of Q_{rot}), Q_T, Q_{el} and Q_{ion} are compared with available data. In this figure the topmost curve represents Q_{TOT} for which there is no comparison.

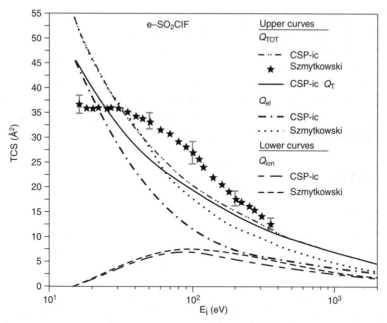

Figure 5.29 Electron scattering with SO_2ClF; *uppermost curves*: Q_{TOT} and Q_T CSP-ic (Joshipura and Gangopadhyay 2008); Q_{TOT}, *star*: measured data (Szmytkowski et al. 2005); Q_{el} *middle curve*: CSP-ic; *dotted curve*: Q_{el} IAM (Szmytkowski et al. 2005); Q_{ion}, *lower curve*: CSP-ic;, *dash*: BEB (Szmytkowski et al. 2005)

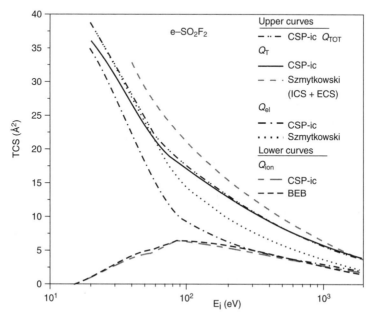

Figure 5.30 Electron scattering with SO_2F_2; *upper curves*: Q_{TOT} and Q_T CSP-ic (Joshipura and Gangopadhyay 2008); Q_T, *uppermost dashed curve (ICS + ECS)*: Szmytkowski et al. (2006), Q_{el}, *chain curve*: CSP-ic; *middle dotted curve*: Szmytkowski et al. (2006); Q_{ion}: *faint dash* CSP-ic; *dash* BEB of Szmytkowski et al. (2006)

Q_T evaluated using CSP-ic are compared with approximate data obtained by Szmytkowski et al. (2006), who calculated the total elastic cross section (ECS) of this molecule using the IAM approximation and the total ionization cross section (ICS) using the BEB theory. Figure 5.30 includes the sum (ICS + ECS) thus obtained by Szmytkowski et al. (2006), to represent Q_T. The IAM data are not expected to be reliable for a large molecule like SO_2F_2 and for intermediate energies it is likely to overestimate the cross section (Figure 5.30). The theoretical results for Q_{ion} are depicted by the lower two curves which are in good mutual accord.

5.4.2 SF_6, H_3PO_4

Sulphur hexafluoride

SF_6 is a non-polar gas that is highly potent as a greenhouse gas but is an excellent electrical insulator. The octahedral SF_6 molecule has a bond length of $R_{S-F} = 1.561$ Å, polarizability 4.49 Å^3 and a rather high ionization threshold 15.32 eV. There have been several theoretical and experimental investigations for e–SF_6 scattering, including the comprehensive recommended data paper by Christophorou and Olthoff (2000). Q_{ion} for this molecule have been measured by Rejoub et al. (2001) and by Christophorou and Olthoff (2004). These data have been included in the electron-scattering review papers relevant to gaseous electronics by Govinda Raju (2011, 2016).

The major total cross sections derived in CSP-ic were reported by Joshipura et al. (2004). The same theoretical approach was employed by Gangopadhyay (2008) for investigating the electron impact ionization of SF_6 along with radicals SF_x ($x = 1$–5). A recent theoretical work on SF_6 extending from low to high energies is from Goswami and Antony (2014).

Total cross sections for e–SF_6 scattering are displayed in Figure 5.31, adopted from Joshipura et al. (2004). Q_T CSP-ic derived cross sections are, in this case, on the higher side of the measured data (Debabneh et al. 1985) up to about 400 eV and are somewhat lower than the AR values of Jiang et al. (1995). Q_{el}, not shown here, were also calculated by Gangopadhyay (2008). Q_{ion} derived by the CSP-ic method Joshipura et al. (2004) agree in general with the rather old but well-known experimental measurements of Rapp and Englander-Golden (1965), and also with the BEB result given in (NIST database). In figure 5.31 the Rapp and Englander-Golden data points shown by inverted triangles are seen clearly after the peak position. It will be noticed that the Q_{ion} peak in SF_6 occurs at a rather high energy around 150 eV. As for the radicals, the CSP-ic Q_{ion} of SF_3 and SF_5 (Gangopadhyay 2008) are much higher than the corresponding experimental data.

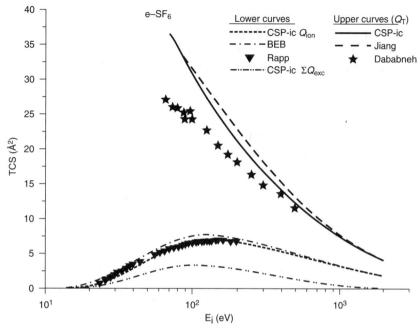

Figure 5.31 Electron scattering from SF_6; Q_T *upper curves:* *continuous:* CSP-ic (Joshipura et al. 2004), *dash:* AR, Jiang et al. (1995); *stars:* experimental data Debabneh et al. (1985); *lower curves:* Q_{ion} *short dash:* CSP-ic, *dash-dot:* BEB (NIST database), experimental Rapp and Englander-Golden (1965); *lowest curve:* ΣQ_{exc}

A numerical summary of the 100 eV TCS of radicals SF_x ($x = 1$–5) and SF_6 from (Gangopadhyay 2008) is provided in Tables 5.6a and 5.6b.

Table 5.6 (a) Ionization cross sections (in Å^2) of SF, SF_2, and SF_3 radicals at 100 eV (a) CSP-ic Gangopadhyay (2008), (b) BEB (NIST database)

SF		SF_2		SF_3	
(a)	*(b)*	*(a)*	*(b)*	*(a)*	*(b)*
4.44	4.44	5.80	5.05	7.94	5.30

Table 5.6 (b) 100 eV cross sections (in Å^2) of SF_4, SF_5, and SF_6; (a) CSP-ic (b) BEB (NIST database), and (c) Christophorou and Olthoff (2004); Q_T and Q_{el} of SF_6 (Gangopadhyay 2008)

SF_4		SF_5		SF_6				
Q_{ion}		Q_{ion}		Q_T	Q_{el}	Q_{ion}		
(a)	*(b)*	*(a)*	*(b)*	–	–	*(a)*	*(b)*	*(c)*
8.61	6.23	8.96	6.81	31.58	21.12	7.28	7.43	6.53

The H_2SO_4 molecule is important in the formation of aerosols in the Earth's atmosphere and has also been observed in planetary environments such as the Jovian satellite Europa. This molecule is polar and tetrahedral with the central S-atom surrounded by four O atoms, of which two are bonded with an H atom. Electron scattering with H_2SO_4

was examined using CSP-ic by Bhowmik et al. (2014). For calculating Q_T and Q_{ion}, the authors, Bhowmik et al. (2014) used a polarizability-based static screening correction, which reduces the group-AR results rather drastically (by ~35%). However due to its aggressive and corrosive nature of this molecule there appear to be no experimental data with which to compare.

Phosphoric acid

H_3PO_4 is a molecule of biological significance due to the similarity of its structure and functional groups with those found in DNA and RNA; in particular H_3PO_4 is of interest as an analogue of the DNA phosphate group.

The molecule is polar and tetrahedral with the central P-atom surrounded by four O atoms, of which three are bonded with an H atom each. Bhowmik (2012) calculated various cross sections including Q_{ion} using the CSP-ic method and the group-AR along with a static screening correction based on polarizability. CSP-ic results and BEB data reported by Mozejko and Sanche (2005) are plotted in Figure 5.32 and a numerical comparison corresponding to $E_i = 100$ eV is also provided in Table 5.7.

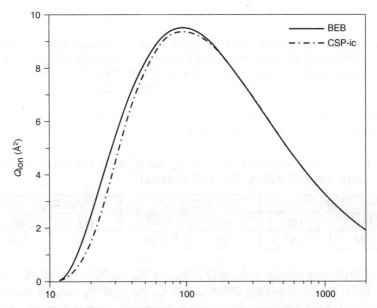

Figure 5.32 Q_{ion} of H_3PO_4 molecule by two different theories

Table 5.7 100 eV theoretical cross sections (in Å^2) for electron scattering from H_3PO_4; (a) Mozejko and Sanche (2005), (b) CSP-ic Bhowmik (2012); Q_T in (a) is the sum of the corresponding Q_{el} and Q_{ion} (in BEB)

Q_T		Q_{el}		Q_{ion}	
(a)	**(b)**	**(a)**	**(b)**	**(a)** **BEB**	**(b)** **CSP-ic**
26.02	24.01	16.51	11.42	9.51	9.37

Apart from the variety of polyatomic targets considered so far in some detail, a quick mention may be made of larger targets exhibiting increasing complexity and structural variety. The greenhouse gas molecule SF_5CF_3 was examined for electron scattering by Vinodkumar et al. (2006) in CSP-ic. Later on the same methodology was also adapted by Kaur and Antony (2015) for C2 to C6 formates, e.g. methyl formate ($HCOOCH_3$), while Goswami and Antony (2014) applied the same theory and calculated cross sections for electron impact ionization of silicon and metal containing organic molecules.

Amongst the still bigger molecular targets and giant clusters, mention must be made of the famous fullerenes in general and C_{60} in particular. Electron impact ionization of C_{60} and C_{70} was investigated experimentally by Worgotter et al. (1994), and theoretically by Deutsch et al. (2000) who found their cross sections to be fairly large. Later, Pal et al. (2011) extended and generalized the modified Jain–Khare semi-empirical formalism to evaluate the partial differential and partial integral ionization cross sections for C_{60} target.

5.5   LARGER HYDROCARBONS AND FLUOROCARBONS

This section will deal with molecules formed by three or four carbon atoms exhibiting a variety of shapes and geometry. Hydrocarbons are of interest due to their occurrence in interstellar medium, and in low-temperature plasmas and in plasma-wall interaction regions.

We begin with, an examination of electron scattering with C3 hydrocarbons, i.e., those having three carbon atoms in each.

5.5.1   C3 hydrocarbons

C_3H_4 (Allene), C_3H_6 (propene), and C_3H_8 (propane)

In a molecule such as C_3H_4 (allene) the C-C bond lengths are larger than the C-H bond lengths and hence the group-AR appears suitable. Total cross sections Q_T for C_3H_4 isomers and as well as for C_3H_8 were measured by Szmytkowski and Kwitnewski (2002) for impact energies ranging from 0.5 to 370 eV. Q_{ion} for C_3H_8 (propane) was experimentally determined by Nishimura and Tawara (1994). Ionization cross sections for C_3H_4, C_3H_6, and C_3H_8 and other hydrocarbon molecules derived using the BEB method are available in the NIST database, while the DM calculations on C_3H_6 isomers were published by Deutsch et al. (2000). For these hydrocarbons CSP-ic calculations using the group-AR were carried out by Vinodkumar et al. (2006). However the derived Q_T of C_3H_4 (not shown here) did not agree with experimental data below 200 eV, and no comparison was given by them for Q_{ion}. Their theoretical Q_T results on C_3H_6 show a similar trend and their Q_{ion} results were in reasonable agreement with experimental data at least in shape. We do not show here the graphical plots but in Table 5.8, the peak Q_{ion} values of three molecules are compared with the BEB NIST data. While there is good agreement for C_3H_8 in the CSP-ic method, the two other cases have significantly higher cross sections than those derived from the BEB method.

Table 5.8 Peak Q_{ion} values (in Å^2) of C_3H_4, C_3H_6, and C_3H_8 at broadly around 80 eV (a) CSP-ic, Vinodkumar et al. (2006); (b) BEB (NIST database)

C_3H_4		C_3H_6		C_3H_8	
(a)	*(b)*	*(a)*	*(b)*	*(a)*	*(b)*
5.83	8.09	7.08	8.74	8.70	8.63

Different total cross sections calculated by Vinodkumar et al. (2006) for the C_3H_8 molecule are reproduced in Figure 5.33 and compared with other data. The CSP-ic Q_T are seen to lie below the measured data especially at lower energies. The CSP-ic Q_{ion} are lower than the measurements of Nishimura and Tawara (1994) but are in good agreement with BEB as well as the earlier experimental data of Schram et al. (1966).

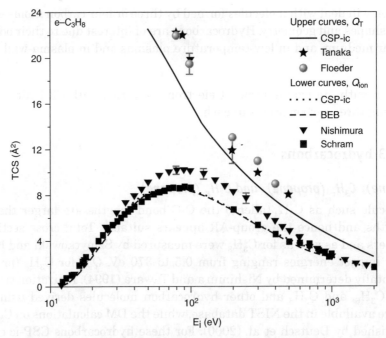

Figure 5.33   TCS of C_3H_8; Q_T *upper curve*: CSP-ic (Vinodkumar et al. 2006); *stars*: measured data, Tanaka et al. (1999); *bullets*: Floeder et al. (1985); Q_{ion} *lower curve*: CSP-ic and BEB (NIST database); *inverted triangle*: measured data, Nishimura and Tawara (1994) and *square*: Schram et al. (1966)

Besides these hydrocarbons, there have also been some electron scattering investigations of C3 fluorocarbons, for example the BEB Q_{ion} for octafluoropropane (C_3F_8) are available in the NIST database and are in good agreement with Q_{ion} derived by Antony et al. (2005) using the CSP-ic method.

5.5.2   C4 molecules

In this subsection we examine electron scattering with C4 molecules, both hydrocarbons and fluorocarbons, relevant in plasma processing.

Open/closed chain isocarbon targets

Of special interest amongst these targets are isomers, i.e., molecules with the same chemical formula but with different structural arrangement of atoms. In this respect, open chain (or acyclic) compounds correspond to molecules with linear rather than cyclic structure. In the following discussion our focus will be on the ionization cross sections since the database for these is more complete.

C_4H_6, C_4F_6

The molecule C_4H_6, may be an open-chain *1,3* butadiene ($CH_2= CH - CH = CH_2$) or *2,*butyne ($CH_3- C \equiv C - CH_3$). CSP-ic calculations using the group-AR with the ionization threshold of *1, 3* butadiene (I = 9.08 eV) were performed by Patel et al. (2014). Figure 5.34 shows Q_{ion} compared with the BEB results of Kim and Irikura (2000) and the experimental data from Kwitnewski et al. (2003) quoted with 20% overall uncertainty. The CSP-ic results differ somewhat from the other two data sets both in shape of the cross section and in its prediction of a lower peak energy value for Q_{ion}.

Figure 5.35 (from Patel et al.) shows Q_{ion} for the other isomer of C_4H_6, *2* butyne (I = 9.58 eV). Once again CSP-ic calculations show lower peak energy than experiment. Unfortunately, there are no BEB data with which to compare. Also note here the work of Szmytkowski et al. (2015) who measured the absolute total cross sections Q_T for the isomer 1, 2-butadiene ($H_2C=C=CHCH_3$) molecule, in the range 0.5–300 eV, and compared their data with their calculations based on the independent atom approximation and BEB cross sections which showed reasonable agreement,

Studying the fluoro-compound C_4F_6, perfluorobutadiene 1,3 ($CF_2= CF - CF= CF_2$) is interesting since its ionization threshold 9.5 eV given in the NIST Chemistry WebBook is significantly different from I = 12.3 eV, as given by Bart et al. (2001). Therefore, Patel et al. (2014) calculated Q_{ion} for this compound with each of these values of 'I' (see figure 5.36). The Q_{ion} of Patel et al. are in a better agreement with the measured data of Bart et al. The maximum Q_{ion} in CSP-ic is 10.46 Å^2 while the BEB (NIST database) value is 10.70 Å^2.

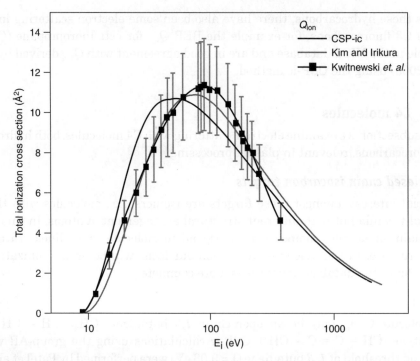

Figure 5.34 Q_{ion} for C_4H_6 (1, 3 butadiene); *continuous curves*: CSP-ic (Patel et al. 2014), BEB (Kim and Irikura 2000); experimental data from Kwitnewski et al. (2003)

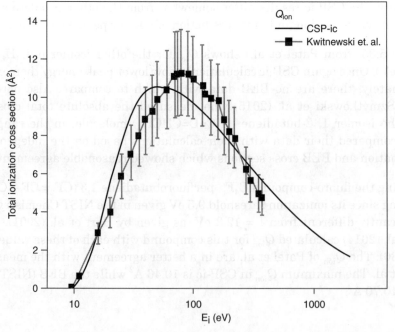

Figure 5.35 Q_{ion} for C_4H_6 (2 butyne); *continuous curve*: CSP-ic (Patel et al. 2014); experimental data from Kwitnewski et al. (2003)

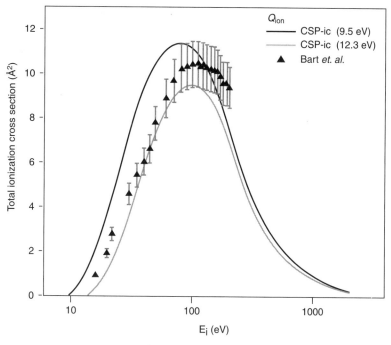

Figure 5.36 Q_{ion} for C_4F_6 (Perfluorobutadiene 1,3); CSP-ic calculations (Patel et al. 2014) with two values of ionization threshold (see text); experimental data – Bart et al. (2001).

C_4H_8, c-C_4H_8, and c-C_4F_8

Moving on to ionization results for the hydrocarbon C_4H_8 i.e. 1 butene $CH_2 = CH - CH_2 - CH_3$, as shown in Figure 5.37, the BEB data of Kim and Irikura (2000) exhibit a lower Q_{ion} peak-value at a slightly higher energy in comparison with the CSP-ic results of Patel et al. (2014). This is partly due to the difference in the first ionization threshold considered in these two theories.

For the isomer *2* butene $CH_3 - CH = CH - CH_3$, values of Q_{ion} are plotted graphically in Figure 5.38. The CSP-ic and BEB theories once again differ somewhat in their peak positions.

All the hydro/fluorocarbon targets considered above are open chain or acyclic molecules. Consider now the closed chain or cyclic compound, cyclobutane or c-C_4H_8 in which the four carbon atoms form a ring-like closed structure but are not coplanar. Ionization cross sections for this and several other cycloalkanes have been studied using the CSP-ic method by Gupta and Antony (2014).

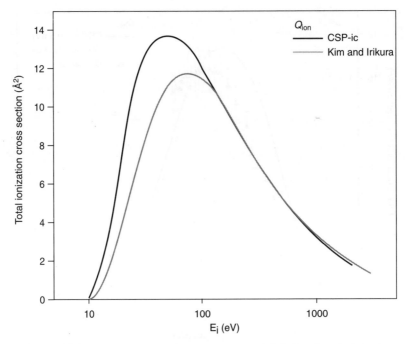

Figure 5.37 Q_{ion} for C_4H_8 1 Butene; *upper curve* CSP-ic Patel et al. (2014); *lower curve* Kim and Irikura (2000)

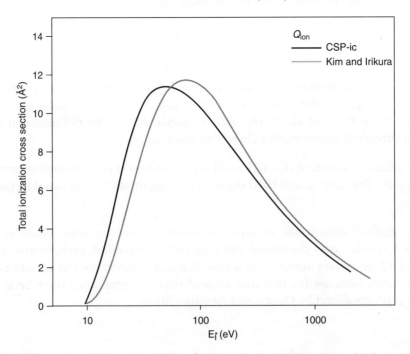

Figure 5.38 Q_{ion} for C_4H_8 2-Butene; CSP-ic (Patel et al. 2014); BEB – Kim and Irikura (2000)

Patel et al. (2014) studied the isomeric effect in the three isomers of C_4H_8 by employing the CSP-ic. Figure 5.39 compares Q_{ion} for the three isomers of C_4H_8. The isomer effect is more evident in closed chain rather than in the open chain molecules and this was reported earlier by Bettega et al. (2006).

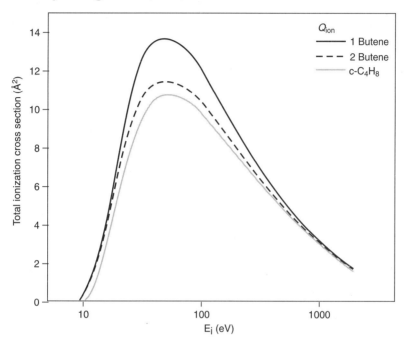

Figure 5.39   CSP-ic ionization cross sections for the three isomers of C_4H_8 (Patel et al. 2014), the *lowest curve* showing c-C_4H_8

The cyclic compound c-C_4F_8 or perfluorocyclobutane, in which the electron impact ionization generates a large number of C_xF_y ionic and neutral species is an important molecule widely used in plasma processing. CSP-ic cross sections for c-C_4F_8 were obtained by Patel et al. (2014). Figure 5.40 shows their calculated results together with other data. The CSP-ic results agree with the other data included in this figure at low and intermediate energies. There is also good agreement with the measured Q_{ion} at least up to the peak. The recommended data (Christophorou and Olthoff 2001) are, however, higher than CSP-ic and the experimental data of Toyoda et al. (1997). The experimental data of Jiao et al. (1998) quote larger uncertainties and are much higher than the CSP-ic results.

Finally it should be noted that theoretical calculations of electron impact ionization cross sections for other cycloalkanes along with aldehyde, and ketone group molecules were undertaken in CSP-ic by Gupta and Antony (2014). Q_{ion} of the famous cyclic compound benzene C_6H_6 has been studied experimentally and also by both BEB Q_{ion} (NIST database) and CSP-ic methods, see Singh et al. (2016).

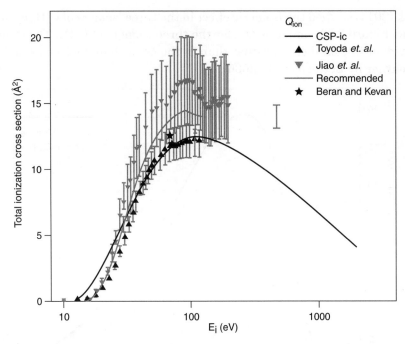

Figure 5.40 Ionization cross sections for c-C_4F_8; *continuous curve*: CSP-ic (Patel et al. 2014); recommended data (Christophorou and Olthoff 2001); *triangle*: experimental data, Toyoda et al. (1997); *inverted triangle*: Jiao et al. (1998); *star*: at 70 eV, Beran and Kevan (1969)

5.6 MOLECULES OF BIOLOGICAL INTEREST

At the beginning of this century, new avenues in electron scattering studies were opened when the role of low energy electrons (1-20 eV) in inducing strand breaks in DNA were recognized (Boudaïffa et al. 2000). This led to a new understanding of radiation damage in cellular systems in which secondary electrons, released by incident primary ionizing radiation, play a significant role (see Chapter 6 for further elaboration). This gave rise to significant increase in the study of electron-induced mechanisms related to the radiation damage of DNA, which includes both direct processes of ionization and electronic-vibrational-rotational excitations, and indirect processes involving resonances in dissociation and dissociative electron attachment (see Mozejko and Sanche 2005, Khakoo et al. 2010). In turn, this led to experimental and theoretical research on large polyatomic molecules of biological interest. Two of the relevant molecules, phosphoric acid and trimethyl phosphate (TMP), have been investigated by Bhowmik (2012), as already mentioned. The present section is devoted to a few other bio-molecular targets for which significant data sets have been acquired.

5.6.1   Furan, tetrahydrofuran, 2,5 dimethyl furan, and beyond

Bernhardt and Paretzke (2003) calculated electron impact ionization cross sections for fragments of DNA, e.g. the bases adenine, cytosine, guanine and thymine, and the sugar-phosphate backbone, by employing DM and the BEB methods. Several bio-compounds have heterocyclic molecules or their derivatives as their subunits. The simplest five-membered heterocyclic molecule with an oxygen atom in the ring is furan (C_4H_4O). Indeed, furan may serve as a prototype for the furanose building unit of bio-molecules and it can serve as a simple analogue of more complex biologically relevant molecules (Mozejko et al. 2012). Thus a study of electron induced reactions with furan is necessary to understand and model electron-driven processes in living cells.

Furan is a pentagonal heterocyclic molecule for which BEB results were given by Szmytkowski et al. (2010). Limbachiya et al. (2015) studied three targets: Furan (C_4H_4O), tetrahydrofuran (C_4H_8O), and 2,5 dimethylfuran (C_6H_8O) employing CSP-ic coupled with group-AR (without screening correction). Limbachiya et al. (2015) also computed Q_{ion} for these targets using the BEB method obtained from the quantemol-N package (Tennyson et al. 2007, Tennyson 2010). Figure 5.41 (Swadia 2017) shows a plot of the CSP-ic Q_{ion} of these three targets together with the experimental measurements of Dampc et al. (2015) for furan. The CSP-ic Q_{ion} results appear high although still within the stated uncertainties of the experiment.

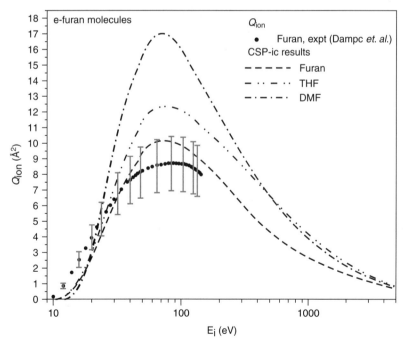

Figure 5.41   CSP-ic Q_{ion} for furan, THF, and DMF along with measured data on furan (Swadia 2017)

The TCS Q_{inel} frequently mentioned in our discussion, is important in determining the energy deposition within the cell, and while it cannot be measured directly it may be estimated by,

$$Q_{inel} = Q_T - Q_{el}$$

with Q_T and Q_{el} being measurable quantities. However the error bars in such data are significant and the above difference may, within these errors bars, even be negative! The complex potential formulation affords a direct calculation of the total inelastic cross section from the complex phase shifts. Fuss et al. (2009), in a joint experimental and theoretical paper, calculated Q_{el} and Q_{inel} for THF (C_4H_8O) target using a complex (optical) potential method with screening corrected Independent Atom representation. Baek et al. (2012) provided the total inelastic cross sections for THF by subtracting the integral elastic cross sections (obtained by integrating the measured differential cross sections) from the measured total cross sections Q_T. Therefore, Limbachiya et al. were able to compare their SCOP Q_{inel} with those of Fuss et al. (2009), as well as Baek et al. (2012). The results could not be reproduced here, but we note that the agreement was fairly good. Further, from the reported Q_{inel} and Q_{ion} of Fuss et al. (2009), a comparative data-set for ΣQ_{exc} for this target was prepared by Limbachiya et al.

Ionization cross sections for THF were derived by Limbachiya et al. and compared with the BEB values of Mozejko and Sanche (2005) and theoretical cross sections in the first Born framework as calculated by Champion (2013). The measured data from Dampc et al. (2011) and Fuss et al. (2009) were also included by Limbachiya et al. There was good agreement between the results of Limbachiya et al., Mozejko and Sanche, and the measurements of Fuss et al., but the theoretical values of Champion and measurements of Dampc et al., though mutually closer, showed a different energy dependence. The electronic-sum obtained by Limbachiya et al. (not shown here) in the CSP-ic appears to be reasonable, while that of Fuss et al. (2009) differs in its energy dependence.

For 2, 5 dimethylfuran (C_6H_8O) or DMF for short, the Q_{inel} derived by Limbachiya et al. (2015) tend to merge with the Q_{ion} at high energies, as expected. There is a good agreement between the Q_{ion} results derived using in CSP-ic, BEB and the measured data (within 18% experimental uncertainties) of Jiao et al. (2009).

Finally in Table 5.9 numerical CSP-ic results at the peak position for ionization and electronic-sum cross sections are shown for five bio-molecules, furan, THF, DMF, H_3PO_4, and TMP. The peak position of Q_{ion} for the first three of these targets is around 70 eV, while for the last two targets the peak occurs around 100 eV.

Table 5.9 Peak Q_{ion} and peak ΣQ_{exc} values in CSP-ic; * from Limbachiya et al. (2015), ** from Bhowmik (2012)

Molecular target	Peak Q_{ion} in Å² (peak energy in eV)	Peak ΣQ_{exc} in Å² Around 40 eV
Furan*	10.14 (70)	4.77
Tetrahydrofuran (THF)*	12.35 (70)	5.81
2,5 dimethylfuran (DMF)*	16.97 (70)	8.22
Phosphoric acid**	09.40 (90)	4.60
Trimethyl phosphate (TMP) **	16.58 (100)	8.17

Studies of electron scattering from even more complex bio-molecules is challenging both experimentally (where it is hard to produce well characterized beam of such targets) and theoretically (due to larger number of electrons and multiple scattering centres). Jones et al. (2016) have carried out experimental and theoretical investigations for integral elastic, electronic-state, ionization, and total cross sections for electron scattering with furfural ($C_5H_4O_2$) molecule. Vinodkumar et al. (2013) explored the ionization of adenine, guanine, cytosine, thymine, uracil, phosphate (H_3PO_4), and DNA-RNA sugar backbone components, by applying CSP-ic along with group-AR without screening. Theoretical total electron scattering cross sections between 0.2 – 6 keV have recently been determined for five nucleic bases together with other bio-molecular targets by Gurung and Ariyasunghe (2017), by using a simple model based on effective atomic total electron scattering cross sections. These papers provide at least an initial estimate of important scattering cross sections but these should not be seen as definitive values.

5.7 GENERAL TRENDS AND CORRELATIONS

Finally once again let us attempt to examine general trends or correlation amongst some of the polyatomic molecules included in this chapter. As a test case, we consider a few tetrahedral targets and also include H_2O and CH_4 for comparison. Our aim is to explore the behaviour of the maximum ionization cross section σ_{max} with respect to the molecular property $P_A = \sqrt{(\alpha_0/I)}$. A graphical plot for a total of 11 molecules is displayed in Figure 5.42.

The linear correlation coefficient $R^2 = 0.846$ is not as good as that shown in Figure 4.40 (Chapter 4) for simpler molecules. However, the correlation in Figure 5.42 improves to $R^2 = 0.977$ after removing SiH_4 and GeH_4 from the linear plot.

Thus, although the peak cross section (σ_{max}) as well as the incident energy (ε_{ion}) at the peak position vary from target to target, some general trends are discernible. As regards the connection between dynamic and static parameters of molecular targets, it has been suggested that the de Broglie wavelength of the incident electron λ_{dB} corresponding to peak energy ε_{ion} is in some kind of resonance with a target property, e.g. a typical bond length of the molecule, see Harland and Vallance (1997), Bull et al. (2012) and references therein, Gupta et al. (2017). Bull et al. (2012) observed a linear empirical correlation between experimental σ_{max} and average polarizability α_0 for a large number of polyatomic organic and halocarbon species. A recent study of this kind reported by Gupta et al. (2017) who have investigated a few C_2F_x ($x = 1 - 6$) and C_3F_x ($x = 1 - 8$) molecules, has revealed a linear correlation between σ_{max} and α_0. A correlation between σ_{max} and $(\alpha_0)^{2/3}$ can also be examined. Correlation studies of this kind provide a general physical feeling of consistency, and can lead to a reasonable estimate of one of the two quantities if the other is known and so allow an estimated cross section to be derived for a target for which no other data exists.

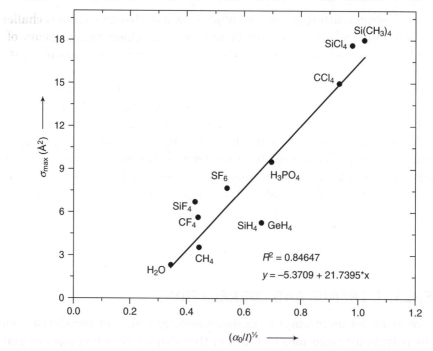

Figure 5.42   Linear correlation of σ_{max} as a function of $\sqrt{(\alpha_0/I)}$, for a set of 11 molecules

With this, we conclude the review of electron scattering from specific polyatomic molecules. While there is plethora of data for electron scattering from a rich diversity of atoms and molecules, these data sets are often incomplete. Semi-empirical and other approximations such as CSP-ic and its variants, the BEB and DM methods, are capable of producing data that is often in good agreement with experiment and may be used to estimate total cross sections that cannot be easily or directly measured (e.g., Q_{exc} or Q_{Ndiss}); but often, there are major discrepancies for which there is no immediate explanation. However, the need for such data is steadily intensifying and so, the next and final chapter reviews the data needs for electron scattering and discusses their application in diverse phenomena, both natural and industrial.

✪ ✪ ✪

6 | APPLICATIONS OF ELECTRON SCATTERING

The previous chapters have highlighted electron induced processes taking place with a wide variety of atoms and molecules, at incident energies from ionization threshold (~15 eV) to about 2000 eV. These studies seek to provide fundamental knowledge and develop insights into these processes that enable us to interpret the relevant phenomena occurring in both natural and technological environments. In this chapter we aim to highlight various applications of such electron scattering data from different atomic–molecular targets. Electrons are almost everywhere in the universe and provide one of the simplest probes for exploring matter in its different forms. Electron collisions with atoms, molecules and ions are dominant in many of the naturally occurring phenomena including the Earth's atmosphere and in the atmospheres of other planets and their satellites, in comets and in far-off molecular clouds of the interstellar medium, where they may play a key role in producing the molecular precursors of life. Primarily the ionosphere of the Earth and other planets is formed by ionization produced by solar UV and X-rays, with some of the photoelectrons produced being energetic enough to cause further ionization along with excitation, leading to the magnificent phenomena of the aurora. The solar wind contains not only electrons (average energy ~12 eV) but protons and other charged particles which produce secondary electrons upon interaction with our upper atmosphere. Furthermore relativistic electrons, though in lower concentrations, are continuously arriving on the Earth as a part of cosmic rays coming from far-off galaxies, etc.

Thus the upper atmospheres of the Earth and planets are a veritable electron collision laboratory in nature. Cross sections for interaction processes of electrons are therefore necessary inputs into the models for understanding physico-chemical and dynamic properties of atmospheres/ionospheres of the Earth and other planets as discussed by Haider et al. (2010, 2012) and others. Energy degradation of electrons resulting from ionization and other inelastic processes in specific atmospheres can be investigated by employing Monte Carlo models as demonstrated in Bhardwaj and Mukundan (2015), and references therein.

Electron scattering discussed in the previous chapters is basically a microscopic, i.e. an atomic–molecular phenomenon, while in applications macroscopic or bulk effects are also important. Therefore, this chapter examines bulk parameters that exhibit dependence on various total cross sections, to describe electron scattering in other phases of matter. One of the most rapidly developing applications of electron scattering is related to radiation damage in biomolecular systems and the response of biological cells to ionizing radiation (Garcia et al. 2015) and therefore a special section is reserved for this topic.

This chapter will also briefly examine interactions between the positron (the anti-particle of the electron) and matter, before we discover to new applications for electron scattering interactions in the coming decades.

6.1 ELECTRON SCATTERING PROCESSES IN NATURE AND TECHNOLOGY

Our understanding of many physical phenomena is strongly coupled to our understanding of the role of electrons in the discrete and continuum excitation of atoms and molecules. Many electron driven processes are observed by the release of energy from an atom or molecule excited by energy transfer from the colliding electrons, energy dissipated in the form of light emitted from the decaying excited state. The most dramatic examples of such electron induced fluorescence are the aurorae. Aurorae are not confined to the Earth but have been observed on Jupiter and Saturn in our own Solar system and are expected to be prominent in many other exoplanetary systems; indeed such phenomena may be one of the first physical processes we may directly observe in exoplanets. An excellent recent review of electron collisions in atmospheres is that of Campbell and Brunger (2016). The 'extreme' case of electron excitation is ionization and this has been central to the discussion in this book. Ionized media or plasmas are the dominant forms of matter in the universe and therefore electron induced ionization is, perhaps, the most important electron collision process. Natural examples include planetary ionospheres, stellar atmospheres and the interstellar medium (ISM). Stellar atmospheres are an example of 'dense plasma' similar to those found in fusion reactors, whereas the ISM is an example of a 'diffuse plasma'. Plasmas are used in many diverse industrial and technological processes. A recent review has highlighted the need to study electron collisions for plasma reactions with biomass (Brunger 2017). Plasma etching has been at the forefront of the development of micro, and now, nanoscale device fabrication. Plasmas also provide one of the main sources of lighting and are increasingly used in medicine with applications from sterilization to surgery.

The prominence of ions in plasmas requires not only electron impact ionization cross sections from neutral species but also data on electron scattering from the product ions, and this is particularly important in dense plasmas where electron scattering from ions may dominate the role of secondary electrons. In contrast to electron scattering from neutral targets, electron–ion collisions are poorly studied, in part, due to the difficulty in preparing high density ion targets. Ionic targets are produced using methods developed by the nuclear and high energy physics communities (Brown 2004). Ions can be produced in almost any charge state from protons to helium-like uranium with the ECR (Electron Cyclotron Resonance) methodology now being most commonly used. Ions extracted from the source are collimated and focussed into ion beams before interacting with the electron beam in a mode known as 'merged beam' with the resultant analysis of scattered electrons and the detection of collision products providing the main experimental knowledge of electron–ion collisions (Schippers 2015). In such experiments total scattering cross sections can be derived, as can excitation and ionization cross sections of the target ion.

The process of Dissociative Recombination (DR) in which the incident electron is captured by the cationic molecular target to form a neutral molecule that subsequently decays to yield often reactive by-products is an important process in many fusion and other dense plasmas but significant differences remain in experimental data and there are often major differences between experiment and theory. These differences may, in part, be due to difficulties in preparing the molecular cations in their ground state. Often, such ions are produced in highly rotationally or vibrationally excited states and in many cases a significant proportion of the ion beam is in metastable excited states. Trapping ions within a storage ring may both allow the density of the ion target to be increased and allow the ions to cool to lower lying states before collision (Guberman 2003) and such storage rings have been used to explore DR for a range of atmospheric (O_2^+, N_2^+) and astronomical targets (H_3^+ and HeH^+). Storage ring experiments have also been used to test the validity of theory but, due to systematic errors that bedevil the experiments, it is generally accepted that theoretical data may provide the best estimate of DR cross sections, even if they are dominated by resonant phenomena.

Electron induced dissociation of molecular targets yields fragments that may strongly influence the local chemistry, for example by the production of radicals or 'energy rich' long lived (metastable) fragments. Such 'electron induced chemistry' (Bohler et al. 2013) may be controlled by 'tuning' the incident electron energy to select specific dissociation pathways, this is particularly true when the process known as Dissociative Electron Attachment (DEA) contributes to the collisional process. DEA is most effective at low energies but recently it has been shown that such chemistry can occur at higher energies where electron impact Dissociative Ionization (DI) is also able to influence the local chemistry (Engmann et al. 2013). DI induced chemistry may be the most important mechanism in Focused Electron Beam Induced Deposition (FEBID) where a high energy (keV) electron beam passes through a beam of precursor gas above a substrate. Electron induced dissociation fragments the FEBID precursor gas to create free metal atoms which subsequently deposit on the surface to build nanostructures. However, DEA and DI cross sections are only available for a few relevant targets and, often only as relative values derived from measured ion yields. Theoretical calculations are currently unable to provide a full description of the branching ratios of DI and DEA.

6.2   ELECTRON SCATTERING IN DIFFERENT PHASES OF MATTER

Throughout this book, references to atomic or molecular targets have been mainly confined to their gaseous state. However, many electron scattering processes in industry and technology occur in other forms or phases. In dense gases, the molecular targets may 'cluster' to form dimers or larger systems, for example, in a humid atmosphere the target molecule AB may cluster with n water molecules to form $AB.(H_2O)_n$. Ionization of such clusters tends to show the product ions A^+/B^+ or A^-/B^- solvated with m water molecules forming $A^+ \cdot (H_2O)_m$ or $B^- \cdot (H_2O)_m$. Electron scattering from such a cluster may be treated as a multiple scattering problem with scattering from each component treated successively. To date this approach has only been developed by a few groups, Bouchiha et al. (2008), Caprasecca et al. (2009) and Fabrikant et al. (2012). More recently, the first full R-matrix calculation of a complex system of two hydrated biomolecules (pyridine and thymine) has been reported by Sieradzka and Gorfinkiel (2017). Experiments investigating electron interactions with clusters have shown that many internal processes (physical and chemical) within the cluster may affect the final products and these are not predicted from simple 'multiple scattering' calculations for example the nucleophilic displacement (S_N2) reaction $F^- + CH_3Cl \rightarrow CH_3F + Cl^-$ may be induced by resonant electron capture in gas phase binary clusters of NF_3 and CH_3Cl (Langer et al. 2000).

Once the target molecules are in bulk or condensed phase the scattering problem becomes a more complex, multidimensional, problem with the energetics of the system being dramatically altered compared to those in the gas phase. For example in the condensed phase the excitation (and hence ionization) energies of the molecule are altered due to interactions with its neighbours. Thus in the case of water the lowest molecular transition is blue-shifted by more than 1 eV (Mason et al. 2006, Hermann and Schwerdtferger 2011). Furthermore, Rydberg states are suppressed in the condensed phase and may be quenched entirely. Shifting of electronic state excitation energies in turn shifts the position and changes the lifetime of scattering resonances and may significantly alter the magnitude of the cross section. For a comprehensive review of how low energy electron scattering changes from gaseous to condensed phase see Carsky and Curik (2012). The threshold of ionization processes and their cross section(s) are also altered in liquid and condensed phase, as discussed below.

6.2.1   From the Micro to Macro Scale

Cross sections Q_{TOT} (or Q_T), Q_{el}, Q_{inel}, Q_{ion} and such, introduced in the previous chapters, are essentially microscopic quantities, i.e. they refer to a single atomic or molecular target, and are typically of atomic/molecular dimensions. Atomic/molecular or micro-processes are the basis of large-scale or bulk behaviour in a medium or an environment. Therefore, the bulk properties or parameters can be expressed suitably in terms of micro-quantities, namely the cross sections. This, the so-called *micro-to-macro* approach, has been amply explored in literature (see Pandya and Joshipura

2012, and references therein). Macro-quantities such as mean free paths and collision frequencies are used in the investigations of atmospheric physics since 1960s and 70s, but the studies in those days were mainly confined to thermal electrons.

One of the simplest macro-quantities is the macroscopic cross section 'Σ' corresponding to a specific atomic/molecular cross section 'Q'. Consider a homogeneous medium composed of a certain atomic/molecular species having N number of particles per unit volume. If $Q_{el}(E_i)$ is the total elastic cross section for incident mono-energetic electrons having energy E_i, the macroscopic 'elastic' cross section is defined as

$$\Sigma_{el} = N \cdot Q_{el}(E_i) \tag{6.1}$$

The particle number-density in the medium is $N = \rho \cdot N_A/M$, where ρ is the mass-density of the medium, M is the corresponding molar mass and N_A is the Avogadro number. Using (6.1), we can define Σ_{inel} for inelastic scattering or Σ_{ion} for ionization and so on. A macroscopic cross section is expressed in (length)$^{-1}$ and it can be employed appropriately to estimate the energy lost by the incident particle per unit path length, in inelastic collisions. The inverse of 'Σ' denotes a typical length called the mean free path (MFP). Thus, the inelastic mean free path (IMFP) Λ_{inel} is defined as follows

$$\Lambda_{inel} = 1/\Sigma_{inel} \tag{6.2}$$

In a particular homogeneous medium, with a given N, the MFP depends on the incident energy. In a laboratory, experiments are performed with mono-energetic beams of electrons and the incident energy can be varied over a wide range from close to zero to keV. In a natural environment, such as an ionosphere, we have a flux of incident electrons with an (often wide) energy distribution. The MFP is the average distance travelled by the incident electron between two successive collisions. In theoretical simulations or modelling at the bulk level, it is customary to randomize the MFP in the spirit of Monte Carlo methods. It is interesting to note that maximum of inelastic (or ionization) cross section Q_{inel} (or Q_{ion}) corresponds to the minimum of the MFP Λ_{inel} (or Λ_{ion}). Moreover, for a composite bulk system i.e. one with several constituents (e.g., a mixture of gases as in the Earth's atmosphere), we are required to define a composite MFP. Thus, if Λ_1, Λ_2, Λ_3, etc., are individual MFPs of constituents 1, 2, 3, etc., then the composite or effective MFP Λ_{eff} is expressed as

$$1/\Lambda_{eff} = 1/\Lambda_1 + 1/\Lambda_2 + 1/\Lambda_3 \tag{6.3}$$

The collision frequency (in s^{-1}) is another macro-quantity of relevance in different applications. If we have mono-energetic electrons incident upon a homogeneous target medium of particle density N, producing ionization with cross section $Q_{ion}(E_i)$, the ionization collision frequency ν_{ic} is defined as follows

$$\nu_{ic} = N \cdot Q_{ion}(E_i) \cdot \upsilon \tag{6.4}$$

In (6.4), v is the speed of the incident electron. If we assume the electron energy distribution to be Maxwellian, and introduce a variable $\varepsilon = E_i/kT$, where k is the Boltzmann constant then, the effective or average ionization collision frequency $<\nu_{ic}>$ is defined as (Itikawa 1971)

$$<\nu_{ic}> = \frac{4}{3\sqrt{\pi}} \int_I^\infty \nu_{ion} \varepsilon^{\frac{3}{2}} e^{-\varepsilon} d\varepsilon \tag{6.5}$$

This quantity can be used to calculate the number of ion-pairs produced per second or ion-production rates (Pandya 2013, Pandya and Joshipura 2014).

Now, in several media the 'standard' electron scattering cross sections discussed in the preceding chapters are expressed in a format more traditionally associated with the chemistry community, i.e., *rate constants*. Often used as a descriptor for plasmas and in atmospheric models, the rate constants integrate the cross sections over the energy range of electrons, defined by an electron temperature. Different rate constants are appropriate for different physical environments, for example in the interstellar medium and other 'cold plasmas' the molecules are often in their ground state or very low-lying states so the state-to-state rate constant as defined below is appropriate. The state-to-state rate is given by integration over state-to-state cross section for a Maxwell distribution of electrons at a given electron temperature T_e. However, in ion mobility experiments the electrons and the molecules are thermalized, so a fully 'thermal rate' is appropriate. The thermalized rate is when the electron and target state distribution have the same temperature T. These can be obtained by convoluting the state-to-state rate with a Boltzmann distribution of target states.

The rate constant in a thermalized gas is commonly defined as

$$k(T) = \int_0^\infty \nu f(v, T) \sigma(v, T) dv \tag{6.6a}$$

where $f(v, T)$ is the normalized distribution of velocities v at a given temperature and $\sigma(v, T)$ is the temperature averaged cross section of the reaction. The function $f(v, T)$ is commonly assumed to be a Maxwellian function (Carsky and Curik 2012) i.e.,

$$f(v) = \sqrt{\left(\frac{m}{2\pi\,kT}\right)^3} 4\pi v^2 \exp\left(-\frac{mv^2}{2\,kT}\right) \tag{6.6b}$$

In equation (6.6b) 'm' is the particle mass and kT is the product of Boltzmann's constant and thermodynamic temperature, although at low pressures this approximation may be questionable. Such an equation may be adapted for electron scattering and thus, if scattering cross sections are known the corresponding rate constants can be evaluated. Figure 6.1 compares the total elastic scattering cross section (denoted here by σ_{elast}) in argon with the derived rate constant, the minimum at low energies being the well-known Ramsauer–Townsend minimum (Franz 2009).

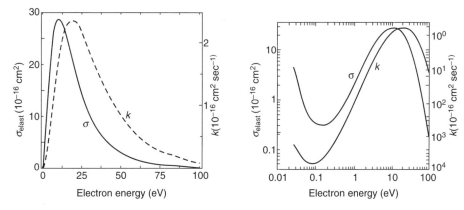

Figure 6.1 Total elastic scattering cross section σ_{elast} for Argon and the corresponding rate constant $k = \sigma_{elast} \times <v_e>$ (Franz 2009)

6.2.2 Modelling Electron Scattering in Solids and Other Condensed Matter

Electron scattering in solids and other condensed matter involves several issues that require careful attention, and it is therefore a field of research in itself. Electron scattering in a variety of solids was examined by Powel and Jablonski (1999), Ziaja et al. (2006), Tanuma et al. (2011), and by others. Tanuma et al. (2011) have reported calculations on electron Inelastic Mean Free Paths (IMFP) in 41 elemental solids, by employing the experimental optical data up to 300 eV, and a simpler single-pole approximation for higher energies. To get an idea of the complexities involved, consider electron interactions in an elemental solid like silicon. The target Si atom is bound, and hence the electron-charge distribution undergoes a significant change compared to a free (unbound) atom. Atomic charge distributions and static potentials for atoms bound in solids were obtained by Salvat and Parellada (1984). A self-consistent field procedure was carried out under appropriate Wigner–Seitz boundary conditions to approximate the atomic charge density $\rho(r)$, which in turn yielded the static potential V_{st} for atom in solid (Salvat and Parellada 1984). Electron exchange and polarization effects may also be included approximately.

Inelastic scattering of electrons in a solid covers a very wide range of phenomena. There are collective or bulk processes such as plasmon excitations, but these can be assumed to be more important at low energies. There are also nearest neighbour effects on the scattering from a particular target. Typically at incident energies above 20 eV the de Broglie wavelength of electron becomes smaller than the usual crystalline bond distances and hence an approximate complex potential single-scattering formulation can be adapted, as in Pandya et al. (2012). Another issue concerns the effective ionization of the target bound in a crystalline solid. In the gas phase, the valence electron is released when the incident electron energy exceeds the first threshold 'I'. In a crystalline medium, the ionization takes place effectively when the incident energy exceeds 'I' by an amount at least equal to the band-gap E_g of the solid. Thus, in the

SCOP formalism, the threshold parameter Δ of the absorption potential V_{abs} (vide previous Chapters 3, 4, 5) must be chosen to be Δ_c, viz.

$$\Delta_c = I + E_g \tag{6.7}$$

The choice of the parameter Δ_c as in equation (6.7) fixes an approximate onset of inelastic processes, and affects the magnitude of the cross section Q_{inel} in a solid. The CSP-ic calculation procedure may be followed for estimating Q_{ion} from Q_{inel}.

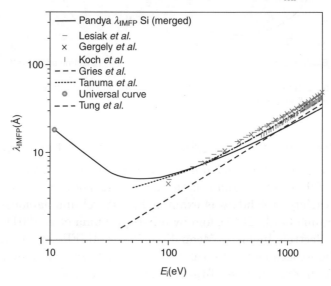

Figure 6.2 Inelastic mean free path λ_{IMFP} of electrons in silicon; *continuous curve*: Pandya et al. (2012) merged smoothly with a universal curve data-point; *bullets*: Rhodin and Gadzuk (1979); *dashes*: Lesiak et al. (1996); *crosses*: Gergely (1997); *dash-dots*: Gries (1996); *vertical line*: Koch (1996); *short dashes*: Tanuma et al. (1988); *dots*: Tung et al. (1979)

In the case of Si (free atom) we have I = 8.15 eV, and in metallic silicon $E_g = 1.1$ eV, so that $\Delta_c = 9.25$ eV; hence the resulting total cross sections using SCOP in the condensed phase turn out to be lower than that of the free atom case. Theoretical descriptions of electron scattering in solids become more approximate compared to free atomic/ molecular targets, in view of the nature of interactions and screening effects of the surrounding atoms/molecules on the target in question.

Calculations for cumulative energy-loss (excluding the bulk) processes of incident electrons in solids were reported by Pandya et al. (2012) to determine the inelastic mean free path (IMFP), denoted by λ_{IMFP}, as a function of energy E_i. The aim of the work was also to find the position of minimum of λ_{IMFP}, and to correlate it with the maximum of the inelastic cross section Q_{inel}. For silicon, we reproduce, in Figure 6.2, the IMFP results of Pandya et al. (2012). Several comparisons have been made here, as mentioned in the figure caption. In this context, the 'universal curve' given by Rhodin and Gadzuk (1979), of which one point is shown here, is the approximate common curve of λ_{IMFP} of all metallic elements. The theoretical results of Pandya et al. show a dip around 65 eV, of minimum magnitude $\lambda_{IMFP} = 5.02$ Å in silicon (Figure 6.2). A point worth noting here is a marked discrepancy amongst the available data at lower energies. However, amongst a variety of inelastic processes, ionization is expected to be dominant for electrons around 65 eV. Pandya et al. also studied the IMFP for solids composed of molecules, e.g. SiO, SiO_2, and Al_2O_3 that exhibit different solid-state structures, and made comparisons with other theoretical data which tend to agree at high energies. Towards lower energies, typically below 100 eV, comparable data are either not available or they differ significantly in energy dependence. Figure 6.3 adopted from Pandya et al. displays mutual comparison of the calculated IMFP for three solids Si, SiO and SiO_2. The minimum IMFP holds significance as it corresponds to the impact energy of maximum inelastic cross section in the solid.

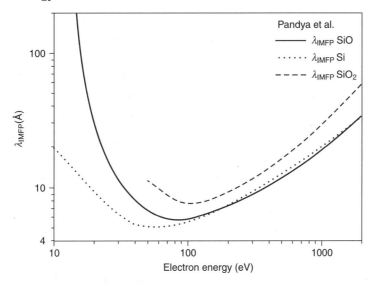

Figure 6.3 Comparison of the calculated IMFP for three solids Si, SiO, and SiO_2 (Pandya et al. 2012)

Electron impact ionization from metallic molybdenum, in comparison with free atomic *Mo*, was studied by Joshi and Joshipura (2015), wherein the CSP-ic methodology was extended to solids by incorporating equation (6.7). Atomic boron and BN in solid phase, relevant in plasma technologies, were investigated by Joshi et al. (2016), who found that the Q_{ion} (per scatterer) of boron in solid state was reduced by about 15% with respect to free B atom around the peak position 60 eV. For boron nitride, Joshi et al. considered three crystalline structures h-BN, c-BN and a-BN, and found that the peak Q_{ion} (at about 50 eV) was the lowest in a-BN.

We turn briefly to a discussion on a special class of solids, which have a multidisciplinary importance.

6.2.3 Electron Interactions in Icy Solids

Icy solids like H_2O-ice, methane-ice, ammonia-ice, dry ice are essentially molecular solids, in that they are composed of molecules held together by intermolecular forces. In the ionization process the embedded molecule is eventually removed from the crystalline structure while the ejected electron(s) as well as the resulting neutral and ionic species produced, in turn, give rise to different chemistries in the medium. As an example of electron scattering calculations consider water-ice which has a rather open tetrahedral structure with four H_2O molecules surrounding a particular molecule. Table 6.1 shows the structure properties of H_2O in the three phases.

Table 6.1 Structural properties of H_2O in three phases (Joshipura et al. 2007, and references therein); * from Timneanu et al. (2004)

Phase	O – H bond length (Å)	O – O bond distance (Å)	I (eV)
Gas	0.96	∞ (free)	12.61
Liquid	0.96	∞ (free)	10.56
Ice	1.01	2.76	11.00 Band-gap (E_g)7.8*

Electron scattering from H_2O in liquid and solid (ice) forms was theoretically investigated by Joshipura et al. (2007) using CSP-ic for single scattering events. In that work, the peak Q_{ion} for H_2O in water ice was found to be 1.59 Å^2 as against the gas-phase value of 2.26 Å^2 for a water molecule (see Chapter 4). Michaud et al. (2003) reported an indirect measurement of the integral ionization cross sections Q_{ion} per scatterer for electrons at 1–100 eV, in an amorphous thin film of ice condensed at 14 K, yielding cross sections considerably lower than that calculated by Joshipura et al. even allowing for the 45% uncertainty in the measurements. We have not shown the Q_{ion} in ice phase as a function of incident energy E_i, but it is found (Joshipura et al. 2007) that in ice, the upper limit of the 100 eV Q_{ion} of Michaud et al. (2003) is close to that of Joshipura et al. In a theoretical work on H_2O-ice, Timneanu et al. (2004) had calculated total inelastic cross sections Q_{inel} for electron scattering in ice and liquid-water from threshold to 1000 eV by employing a semi-empirical optical data model. Specifically, Timneanu et al. used experimental data up to 30 eV, beyond which an extrapolation based on liquid-water, was adopted. Joshipura et al. extracted the theoretical ionization contribution viz., Q_{ion} from Q_{inel} of Timneanu et al., and found that the Q_{ion} (ice) resulting from the Timneanu et al. data were somewhat higher than the gas-phase Q_{ion}.

The electron mean free paths Λ_{inel} and Λ_{ion} are exhibited along with comparisons in Figure 6.4 for liquid-water and ice phases. The two topmost curves in this figure represent the CSP-ic results of Λ_{ion} for liquid-water and ice respectively, followed by the corresponding results of Λ_{inel}. The results from Timneanu et al. (2004) are also included for comparison. All the CSP-ic results (Joshipura et al. 2007) in this figure show a broad minimum around 100 eV. In the ice phase, the minimum Λ_{ion} in CSP-ic is about 20 Å, while it is about 12 Å in the results of Timneanu et al. This reflects the fact that the Q_{ion} (ice) of Timneanu et al. are on the higher side. The difference is also due to approximate nature of calculations in Joshipura et al., which need to be repeated with another value of the input R_p. The dip in MFP corresponds to peak in the corresponding cross section.

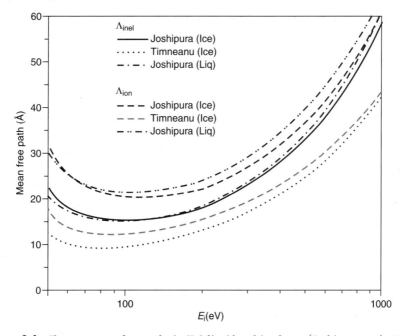

Figure 6.4 Electron mean free paths in H_2O liquid and ice forms (Joshipura et al. 2007)

Laurent et al. (2007) discussed how the gas-phase R-matrix theory could be efficiently extended to carry out calculations in a condensed matter environment, by considering electron interactions in H_2O-ice. With this development, we can expect a better representation of electron scattering in solids and liquids especially at lower energies.

Now consider CO_2-ice or dry ice, which is significant in view of its occurrence on Mars–a planet being extensively explored by international space missions from NASA, ESA and ISRO, India. Pandya (2013) computed CSP-ic values with CO_2 as dry ice wherein the effective ionization threshold I_{th} = 16 eV (Siller et al. 2003), is higher than the first ionization energy I = 13.77 eV of a free or isolated CO_2 molecule. Therefore, the absorption potential, the relevant energy parameter Δ, which indicates the onset of inelastic (ionization) processes, was set to be equal to I_{th}. For dry ice the maximum Q_{ion} = 2.74 Å^2 obtained in the CSP-ic calculations (Pandya 2013), was found to be lower than the gas-phase value 3.70 Å^2 of the CO_2 molecule (Chapter 4). Moreover, a recent determination of the mass-density of dry ice under Martian conditions (Mangan et al.

2017) yields a value 1.64 gm/cc. Thus, the calculated minimum ionization mean free path of electrons in dry ice is about 16 Å, and the minimum takes place broadly around 150 eV.

6.3   THE TERRESTRIAL ATMOSPHERE

One of the first applications of our understanding of electron collisions was to model the Earth's ionosphere. The terrestrial ionosphere was first characterized by the development of radar in the 1930s, whereupon the plasma nature of the Earth's upper atmosphere was identified and it was soon proven that aurorae were the result of electron collisional phenomena.

Subsequently, the phenomena of day and night glow were also discovered and the role of electron interactions with terrestrial atmospheric atoms and molecules quantified. Figure 6.5 shows the structure of the Earth's upper atmosphere and typical electron densities. The ionosphere is formed by interaction of the terrestrial atmosphere with the solar radiation and solar wind, and its persistence and spatial properties are defined by the terrestrial magnetic field. During a solar storm 'coronal ejections' may send a high intensity pulse of charged particles (protons and electrons) into the Earth's upper atmosphere which may produce an avalanche of secondary electrons, that collide with nascent oxygen and nitrogen atoms and molecules, which in turn are excited before decaying, yielding the bright colour curtains often seen in northern and southern polar regions. These are the famous aurorae.

Some of the earliest theoretical estimates of electron scattering cross sections and in particular ionization cross sections were derived to provide input into models and simulations of the Earth's ionosphere and aurorae, with Mott and Massey reporting in their early papers the cross sections for electron scattering from oxygen and nitrogen (Mott and Massey 1965). Experiments to measure total scattering cross sections of molecular oxygen and nitrogen were amongst the first measured (Ramsauer 1921) but studies of electron interactions with atomic oxygen and nitrogen had to wait until the 1960s when beams of O and N atoms could be prepared using microwave discharges (Sunshine et al. 1967, Smith et al. 1962). However it is still difficult to prepare high density pure beams of such atomic targets (>25% O and >10% N), so experimental data for such cross sections remain sparse and models must depend on theoretical derivations. The R-matrix scattering code is now capable of providing highly reliable and accurate (within 10%) cross sections for electron scattering at low energies (below ionization energy) whilst the semi-empirical methods discussed in Chapters 3 to 5 are valid above the ionization threshold. Ionization cross sections have been reliably calculated (see Chapter 3 Section 3.2 and Chapter 4 Section 4.1, 4.2) for all important atoms and molecules in the terrestrial ionosphere.

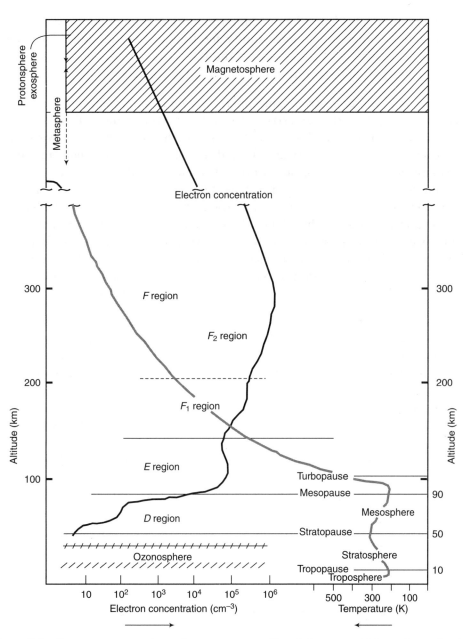

Figure 6.5 Electron concentration and temperature profiles of the terrestrial ionosphere, showing its different regions

The terrestrial ionosphere is divided into regions as shown in Figure 6.5. The E region is unusual in being a region where negative ions (anions) are prevalent. The O^- anion is formed by Dissociative Electron Attachment (DEA) and until recently it was presumed to be unique in the solar system but recent data from the Cassini Mission has shown that the Saturnian moon Titan has a similar layer (see below). The presence of high density of cations provides the opportunity for Dissociative Recombination (DR), a process that is known to be important in the terrestrial atmosphere as a source of emissions and as part of the airglow (Geppert and Larsson 2008).

Thus models of the terrestrial atmosphere rely on a comprehensive knowledge of electron scattering from the terrestrial atoms (O, N, H) and molecules (O_2, N_2, H_2O and CO_2), and we have comprehensive data sets of such cross sections. However, many of these compilations are now rather old and need updating with more recent data. Furthermore, most compilations do not present recommended values and therefore the models end up using different and inconsistent data sets. Accordingly it is timely and necessary to assemble updated data compilations which present recommended values (with uncertainties) that can be adopted by the modelling/simulation community. An excellent example of the assembly and exploitation of such a database is the work of Campbell and Brunger (2007) on electron interactions with NO to determine the electron impact contribution to infrared NO emissions in the aurora and the work of Ashrafi et al. (2009) on nitrogen aurora using electron–N_2 tabulated data.

There is one additional molecule that plays an important role in the Earth's atmosphere: ozone. By absorbing solar UV light, ozone protects the Earth's surface from such mutagenic radiation but it is formed in a limited region known as the stratosphere. Recent studies have shown that during major solar storms and auroral events, secondary electrons may penetrate into the stratosphere; hence a dataset of electron interactions with ozone would be useful. Several experimental studies of electron interactions with ozone have been reported producing data with which to benchmark theory (see Chapter 4) and, in particular, there have been some extensive studies for DEA to ozone yielding of O^- and O_2^- anions. However to date there are no cross sections for the excitation of electronic states of ozone and hence no data on the dissociative fragments produced from such states, while there is only sparse data for vibrational excitation. More experiments and further calculations on electron scattering from ozone are therefore required but, being a highly reactive gas, ozone poses challenges to the experimentalist whilst its low lying electronic states also challenge theoretical calculations.

6.4 THE ROLE OF ELECTRON COLLISIONS IN PLANETARY ATMOSPHERES AND COMETS

6.4.1 Planetary Aurorae

Most of the solar system planets are subject to the same intense bursts from the Sun during the solar cycle and all are subject to similar photoionization processes by solar irradiation. It is therefore not surprising that aurorae were both predicted and later observed on other planets (Figure 6.6a-c) with magnetospheres, such as the gas giant planets (Jupiter, Saturn, Uranus and Neptune). Figure 6.6(b) illustrates an auroral oval observed on Jupiter which is much larger in its extent than those on Earth, Figure 6.6(a), and is caused by emission from a more exotic molecular species H_3^+ and hydrocarbon cations (Gérard et al. 2014).

Figure 6.6(a)   Pictures of aurora: Earth (from https://pl.pinterest.com/pin/312859505342271390/?lp=true)

Figure 6.6(b)   Pictures of aurora: Jupiter (from https://www.nasa.gov/feature/goddard/2016/ hubble-captures-vivid-auroras-in-jupiter- s-atmosphere)

Figure 6.6(c)   Pictures of aurora: Saturn (https://commons.wikimedia.org/wiki/Aurora#Other_Planets)

Aurorae have also been observed on Saturn in both the UV, due to H Lyman-α and H_2 Lyman- and Werner-band transitions, and in the infra-red, arising from H_3^+. However the large gas giants (Jupiter and Saturn) may also have an 'internal' aurora created by ejection of material from their moons which, when ionized by solar radiation, penetrate into the planet's atmosphere causing local aurorae. Indeed such a process derived from the Jovian moon Io leads to permanent UV and X-ray aurora seen on Jupiter while Enceladus, with its water plumes, injects electrons and ions 240,000 km along magnetic field lines into the Saturn's North Pole inducing observed aurorae. Several planetary moons may have their own aurorae, for example the Jovian moon Ganymede, which is also embedded in Jupiter's large magnetosphere, has its own aurorae arising from its own magnetic field believed to be induced by a salt water ocean under ice covered surface. Aurorae have also been observed on planets with no magnetospheres, for example on Mars where the solar wind strikes the planet's atmosphere directly and high energy electrons penetrate deep into Mars' atmosphere exciting carbon dioxide to produce UV aurorae. Weak emissions observed in the atmosphere of Venus have also been ascribed to auroral phenomena.

Thus the study of planetary aurorae requires extensive knowledge of electron impact with many types of molecular systems ranging from from simple hydrogen to hydrocarbon species e.g. methane, acetylene, ethylene and ethane (Perry et al. 1999). Much of the data required are the higher energy cross sections for ionization described in this book but dissociative recombination cross sections are also important, for example, for H_3^+ and several hydrocarbon cations that play important roles in the Jovian system. Once again, databases for molecules believed to contribute to planetary studies have been assembled but are often incomplete and different models have adopted different data sets. The accumulation of recent data from Cassini mission for Saturnian aurorae, new data on Martian aurora by the Maven-mission instrument on the Exo-Mars Trace Gas Orbiter as well as new data expected from the Jovian system by Juno and later Juice missions, all require fresh collation of such data for electrons from 10 eV to 100 keV. High energy data may use analytical formulae for derivations and extrapolations at higher energies (>1 keV), but there is still need for both experimental and theoretical effort across this wide energy regime.

6.4.2 Planetary Ionospheres

The study of planetary aurorae is intrinsically linked to the physical and chemical structure of the planetary ionosphere. Detailed models of the Jovian and Saturnian ionospheres have been developed in the last decades. For example, for Jupiter, models were developed to interpret mission data (Grodent et al. 2001, Millward 2002). Similarly, the Thermosphere–Ionosphere Model (STIM) was developed to quantify observed aurora on Saturn by Garland et al. (2011). Such models contain extensive electron scattering cross sections over the relevant wide energy range, as mentioned above. In the terrestrial atmosphere, one particular region, the so-called E-region, is where observations indicate that anions are formed. This is ascribed to the formation of O^-

anions by Dissociative Electron Attachment to molecular oxygen. Until recently this was the only known case of anion formation in any planetary atmosphere, but in 2004 the Cassini–Huygens mission reached the Saturnian system and explored the atmosphere of the Saturnian moon Titan, which is the only solar system moon to have a thick atmosphere, composed of 98.4% nitrogen and 1.4% methane, with a surface pressure in excess of that of Earth at 1.5 bars. The Electron Spectrometer (ELS), one of the sensors making up the Cassini Plasma Spectrometer (CAPS), revealed the existence of copious amounts of negative ions in Titan's upper atmosphere (Coates et al. 2007). The observations at closest approach (~1000 km) show evidence for negatively charged ions up to ~10,000 amu/q, as well as two distinct peaks at 22±4 and 44±8 amu/q with evidence for a third one at 82±14 amu/q (Figure 6.7). The formation of these anions may be attributed to dissociative electron attachment to a set of molecules known as tholins, a class of heteropolymer molecules formed by solar ultraviolet irradiation of methane and nitrogen, the simplest building blocks of which are HCN and CH_3N. The negative charge may then be transferred to more acidic molecules such as HC_3N, HC_5N or C_4H_2. The three low mass peaks observed by the ELS are then attributed to CN^-, C_3N^- and C_5N^- which can then act as intermediates in the formation of the even larger negative ions observed by ELS, which are in turn precursors to the aerosols observed at lower altitudes. Chemical and physical models of Titan's ionosphere have been developed by Vuitton and co-workers (2009) to interpret this data and require further data on electron interactions with nitriles and tholin compounds. This research is given further emphasis by the discovery of tholins on Pluto and its moon Charon by the New Horizons mission requiring updated and more detailed models of the solar wind interaction with Pluto's atmosphere.

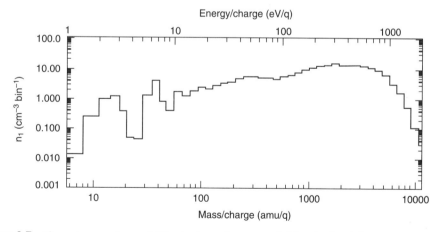

Figure 6.7 The anion spectrum of Titan's atmosphere recorded by the Cassini-Huygens mission at an altitude of 953 km (Coates et al. 2007)

Now, in an attempt to develop a quantitative feeling about the order of magnitudes of the macro-parameters involved in the electron collisions in various gaseous environments highlighted above, we explore some of the calculated results through graphical presentations.

6.4.3 Electron Mean Free Path

As an illustration, consider the calculations of MFP in the Earth's ionosphere, which require cross sections and other inputs viz., concentration (i.e. number-density) of major species N_2, followed by O_2, Ar, O, etc., at different altitudes for which the relevant data is available from Hedin (1991).

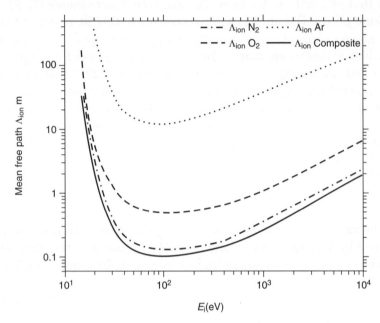

Figure 6.8   MFPs Λ_{ion} (in m) our ionosphere at 80 km (Lat. 70^0 N, Long. 100^0 W); from Pandya et al. (2010)

Figure 6.8, reproduced from (Pandya 2013, Pandya and Joshipura 2014), displays the plot of individual as well as composite ionization MFPs Λ_{ion} (in metres) versus electron energy, at an altitude of 80 km in our ionosphere. As can be seen in this figure the composite MFP is quite close to the Λ_{ion} of the dominant gas N_2, as expected, and it exhibits a minimum around 90 eV, at which energy Q_{ion} attains the maximum. The magnitudes of MFP depend on the type and concentrations of the dominant species, which in turn depends on the altitude.

6.4.4 Collision Frequency

In this discussion we focus first on a relatively low energy phenomenon viz., neutral dissociation upon electron impact (see Chapters 2 and 4), and highlight calculations for the atmosphere of the H_2-rich Jupiter. Relevant inputs are needed to obtain theoretical collision frequency $<v_{ic}>$ defined in equation (6.5), corresponding to neutral dissociation of $H_2(X^1\Sigma_g^+)$ into two neutral H(1s) atoms through the excitation to $b^3\Sigma_u^+$. Total neutral dissociation cross sections of molecular hydrogen are already discussed in Chapter 2, as a part of the discourse on e–H_2 scattering. In estimating $<v_{ic}>$, the known number densities of H_2 at the typical altitudes of 50 and 200 km in the Jovian atmosphere were employed (Pandya et al. 2010; Pandya 2013; Pandya and Joshipura 2010, 2014).

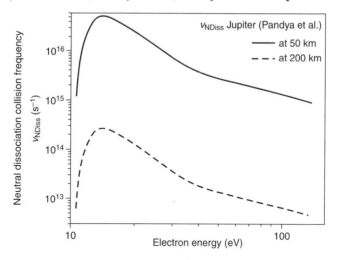

Figure 6.9   Effective collision frequency $v_{\text{N-diss}}$ (in s^{-1}) for neutral dissociation of H_2 in Jovian atmosphere (Pandya 2013)

Neutral dissociative collision frequencies $v_{\text{N-diss}}(s^{-1})$ as functions of electron energy E_i at these altitudes are shown in Figure 6.9, reproduced from Pandya (2013). The collision frequency reaches the peak near the maximum of the TCS Q_{Ndiss} (see Chapter 2) i.e. around 15 eV. The difference in the magnitudes of $v_{\text{N-diss}}$ in the two curves arises due to the difference in the number densities of H_2 at these altitudes.

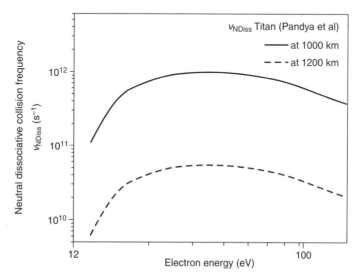

Figure 6.10   Effective collision frequency $v_{\text{N-diss}}$ (in s^{-1}) for neutral dissociation of N_2 on Titan, at two altitudes (Pandya et al. 2010)

Consider Titan, the largest Saturnian satellite, which has an atmosphere rich in N_2, having a concentration of ~98% in its stratosphere. Figure 6.10, adopted from Pandya (2013), also Pandya et al. (2010), depicts the effective neutral dissociative collision frequency $\nu_{N\text{-diss}}$ for N_2 at two altitudes on this Saturnian satellite. The distribution of effective collision frequency for N_2 on Titan shows a broad peak as against a sharp profile for H_2 on Jupiter as shown in Figure 6.9. This difference arises due to the dissociative ionization. For H_2 the dissociative ionization has a very small contribution (<1%); whereas for N_2 there is a significant contribution from dissociative ionization above 30 eV, which effectively broadens the distribution (Figure 6.10).

In the Earth's upper atmosphere the so-called precipitating electrons produce ionization at high latitudes. The ionizing collision frequency in this case was calculated by Pandya (2013). Figure 6.11 (Pandya 2013) exhibits the ionizing collision frequency ν_{ic} (in s^{-1}) as a function of electron temperature for nitrogen molecules in the Earth's ionosphere (Lat. 70°N, Long. 100°W) at two altitudes, namely, 80 and 100 km. The Q_{ion} of N_2 calculated using CSP-ic (Chapter 4) along with other input data were employed to obtain these results. The upper curve showing ν_{ic} in the ionospheric D region peaks broadly at the value 1.5×10^7 s^{-1} for incident electrons of about 150 eV energy. The difference in the results at the two altitudes is attributed to the number of N_2 molecules available at these locations. Other relevant quantities such as ion production rates (in cm^{-3}s^{-1}) can also be calculated by starting with the Q_{ion} data and other relevant inputs (Pandya 2013).

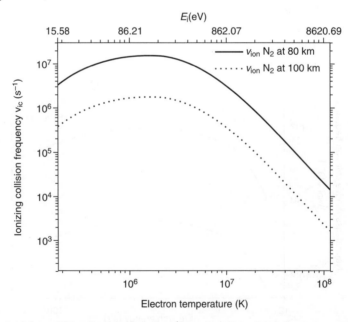

Figure 6.11 Ionizing collision frequency ν_{ic} (in s^{-1}) as a function of electron energy and temperature, for N_2 in the earth's ionosphere (Lat. 70° N, Long. 100° W); upper curve 80 km, lower curve 100 km

6.4.5 Electron Collisions in Comets

One of the unexpected yet fascinating discoveries of the recent ESA Rosetta mission to Comet 67P/Churyumov–Gerasimenko has been the recognition that electrons, not photons as previously assumed, were responsible for the dissociation of water and carbon dioxide erupting from the comet surface as it approached the Sun resulting in the emission spectra recorded by the ALICE spectrometer in the UV and OSIRIS camera (Feldman et al. 2015, 2016; Bodewits et al. 2016). In particular, the observed OH emission can be attributed to electron dissociative excitation of H_2O while most of the 630 nm emission from [O] in the inner coma of the comet arises from electron impact dissociative excitation of CO_2. The emissions detected in the NH filter of the OSIRIS camera are most likely the result of electron dissociative excitation that produces OH^+ ions from H_2O molecules directly in an excited state, while data from the CN filter suggest that further emissions are the result of CO_2^+ ions produced by electron impact dissociative excitation and of fluorescent emission by CN, radicals most likely formed from dissociation of the HCN molecule. However, the lack of experimental cross sections hampers the interpretation of these observations, and cross sections for the production of $O(^1D)$ from CO_2 and H_2O, and of CN from HCN by electron impact are urgently needed.

Electron collisions may also explain many physical phenomena displayed by the comet. The intensity of the emission in the inner coma was seen to decrease, despite increasing gas production rates, as the comet approached the Sun. The most likely explanation is that, because the gas production rates increased much faster than the ionizing solar radiation, collisions between electrons and neutral water molecules lowered electron temperatures in the inner coma below the activation threshold of the dissociative impact excitation reactions. These observations provide the first evidence of discrete electron collisions in an astronomical object and are leading to new programs to explore electron driven processes in other main body comets from ground based observations. For example, electron impact excitation may also explain large-scale jet-like structures around comets (A'Hearn et al. 1986). New databases will need to be assembled to develop models of cometary emissions that include photo-ionization (the source of active electrons); electron induced dissociation yielding fluorescent atomic, molecular and ionic fragments; and electron-driven reactions induced with molecular positive ions including dissociative recombination reactions. The rich inventory of species observed on Comet 67P further complicates the problem since not only water and CO_2 and their fragments need to be included but many hydrocarbons and nitriles must also be included. Key species may be radicals such as OH and CN for which experiments are impractical. Theoretical cross sections for reactive species like OH and CN have been discussed in Chapter 4. Comet 67P data also suggests that the range of electron energies necessary to be studied is large: from the cold component observed below 0.1 eV, to a 5 to 10 eV electron population being interpreted as electrons retaining the energy they obtained when released in the ionization process, to energies as high as 100–200 eV.

6.5 THE ROLE OF ELECTRONS IN ASTROPHYSICS AND ASTROCHEMISTRY

6.5.1 Astrophysics and the Early Universe

In the beginning, just one second after the 'big bang' the universe was filled with neutrons, protons, electrons, positrons, photons and neutrinos. Within three minutes the first elements were formed during a process known as Big Bang nucleosynthesis with protons and neutrons colliding to make deuterium, most of which combined to make helium and trace amounts of lithium. Hydrogen and helium atoms then began to form as the density of the universe fell, and this is thought to have occurred some 377,000 years after the Big Bang. Initially hydrogen atoms were ionized but as the universe cooled down, electrons were captured by the ions in a process called electron recombination. At the end of 'recombination' most of the protons in the universe were bound up in neutral atoms. Thus electron collisions are amongst the first and most important collisions in the history of the universe. Indeed the earliest electron collision processes were electron–proton ($p + e \rightarrow H(n) + h\nu$) and electron–$He_2^+$/$He^+$ recombination ($He^{2+} + e \rightarrow He^+ + h\nu$/ $He^+ + e \rightarrow He + h\nu$). Before recombination all the electrons were free; after recombination, the electrons and protons were mostly bound to form atoms. Similarly before such recombination occurred, most of the photons (light) in the universe were interacting with electrons (and protons) through Thomson scattering, as the mean free path each photon could travel before encountering an electron was very short and so the universe was opaque. But once most of the protons in the universe were bound, the photons' mean free path became effectively infinite and the photons could now travel freely across the universe which had now become 'transparent'. This cosmic event is usually referred to as decoupling. The photons present at the time of decoupling are the same photons that we see in the cosmic microwave background (CMB) radiation, after being significantly cooled by the expansion of the universe.

Evaluations of the electron–proton recombination cross sections $p + e \rightarrow H + h\nu$ have been extensive with both 'complete' and semi-analytical results being widely reported and reviewed (Janev et al. 2012) for each final state 'n' of the product H atom (H{n}) from 0.1 to 10^4 eV. Similarly for electron–He^{2+}/He^+ concise and complete data sets have been measured and evaluated. A small amount of lithium was also created in the early stages of the universe via

$$p + {}^4He \rightarrow {}^5Li + h\nu$$

This process requires electron interactions with Li cations as an initial step.

6.5.2 Astrochemistry in the Early Universe

The first molecules to form in the universe are believed to be H_2, D_2 D_2^+, H_3^+, H_2D^+, D_2H^+, D_3^+, HeH^+, HeD^+, He_2^+, LiH^+, LiD^+, LiD, LiH^+ (Lepp et al. 2002), most of which were formed in ion molecule collisions which are 'barrierless' with rate constants that are often increasing at lower temperatures. The first electron–molecule collisions were therefore once again recombination reactions such as

$$H_3^+ + e \rightarrow H + H + H \rightarrow H_2 + H$$

and

$$e + HeH^+ \rightarrow He + H$$

Such dissociative recombination reactions have been the subject of considerable debate within the atomic and molecular community since experiments and theoretical estimates of the cross sections (rate constants) have revealed orders of magnitude differences. Experiments performed in ion storage rings have, in part, resolved such differences by revealing a strong dependence on the internal (ro-vibrational) energy of the target ion. By cooling the ions in 'storage rings' before colliding the ions with electrons, the experimental data was found to be in a better agreement with theoretical data which assumed the target to be in its lowest state of excitation (Larsson 2012).

An additional electron process leading to the first anions in the universe is electron attachment

$$e + H \rightarrow H^- + h\nu$$

which provided a route to the formation of molecular hydrogen

$$H^- + H \rightarrow H_2 + e$$

by associative detachment. Indeed, this is the most probable route by which early H_2 was formed in the universe before the first stars and the subsequent production of 'dust', the surface of which is now believed to be the major source for molecular hydrogen production (see below).

6.5.3   Astrochemistry in the Interstellar Medium

Once the first stars had formed and died yielding 'dust', surface chemistry became possible and from this, larger and more complex molecules could be formed including those prebiotic molecules that were the harbingers of life. Today the study of molecular formation in space is an active astronomical discipline commonly known as 'Astrochemistry', which in turn is a part of the developing interdisciplinary field of 'Astrobiology', and the study of the origins of life on Earth and prospects for life having developed in other solar systems with their 'exoplanets'. Figure 6.12 shows schematically how such dust grain surface chemistry is believed to lead to synthesis of molecular species in the Inter-Stellar Medium (ISM): those regions between the stars in any galaxy which contain 'dust clouds' from which stars and planets from ISM.

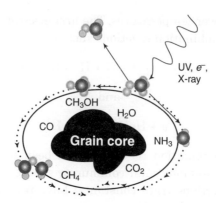

Figure 6.12 A schematic diagram of molecular synthesis on dust grains in the Inter-Stellar Medium

Dust grains comprize a core of carbon or silicate with a mantle of condensed 'primary' molecules such as CO and CO_2 formed by ion molecule reactions. Such grain surfaces act as a reservoir upon which H atoms recombine to form molecular hydrogen, the most common molecular species in the universe. The mantle may be 'processed' by energetic cosmic rays traversing the ISM as well as UV photons and X-rays emitted from stars. Until recently UV processing was considered to be the major process for such mantles but recent astrochemical studies have shown that neutrals (e.g., H, O and N atoms) play an important role in synthesis of several simple molecules in the mantle including water, methane, methanol, ethanol, ammonia and formaldehyde.

When any ionizing radiation (UV photons, X-rays or cosmic rays, protons and He^+) interact with the grain core or the ice mantles, large numbers of secondary electrons will be liberated, with ~4×10^4 electrons per MeV of energy deposited. Until recently the role of these electrons in processing of the ice and in desorbing species from the mantle into the gas phase has been neglected but recent experiments have shown that they may be important, if not essential, in the production of complex organic molecules (COMs), as examined by Mason et al. (2014) and Boyer et al. (2016). By early 2017, over 180 molecular species had been observed in the ISM and many more are expected with the commissioning of the ALMA telescope and launch of the James Clark Maxwell space telescope; with many of these molecules likely to be synthesized by electron processing of ISM ices. Experimental studies of electron induced chemistry in ISM ices are still at an early stage exploring the variety of molecules that can be synthesized in different ices-pure, binary and more complex. For instance, Mason and co-workers have explored molecular synthesis in pure oxygen (Sivaraman et al. 2011) and nitrous oxide ices (Sivaraman et al. 2008), and binary mixtures of both ammonia and methanol, and ammonia and carbon dioxide ices (Jheeta et al. 2013, 2012), whilst Arumainayagam and co- workers have shown the formation of 15 products in electron irradiation of methanol ices (Boyer et al. 2016). There have been fewer studies exploring the routes by which such molecules are synthesized, defining the underpinning electron processes. One of these, is the study of ozone formation in pure oxygen ice, explored using $^{32}O_2/^{36}O_2$ as reported by Sivaraman et al. (2011). Low energy electron processes such as Dissociative Electron Attachment have been shown to provide pathways for molecular synthesis while at high energies reactive neutral and ionic species released by electron impact induced dissociation, dissociative ionization and ion pair formation may lead to new routes of synthesis with different electron induced processes being dominant at different electron energies, as explained in Figure 6.13.

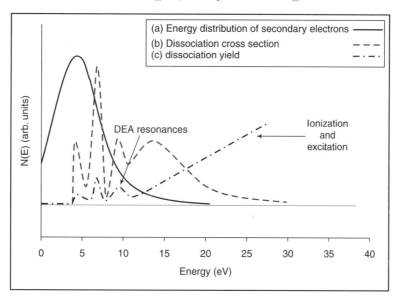

Figure 6.13   Schematic diagram showing the expected secondary electron energy distribution from cosmic ray impact on a dust grain compared with typical product yields induced by dissociative ionization and dissociative excitation and the resonant driven process of Dissociative Electron Attachment (DEA), Boyer et al. (2016)

Comparisons between product yields induced by electrons and UV photons of the same energy are revealing higher yields induced by electrons, once again highlighting the importance of the electron processes. The reason for such differences is not yet fully established: the greater penetration depth of electrons provides some explanation but even allowing for this the electron induced yields are larger. Electrons can excite electronic states 'forbidden' by single photon absorption and thus open new dissociation pathways and these so-called 'dark states' may provide an explanation for greater electron induced yields. Comparing gas phase to condensed phase, both the cross sections for electron scattering and the molecular spectroscopy may be significantly altered. In the condensed phase the threshold for electronic excitation and ionization may be considerably shifted compared to the gas phase with some states (Rydberg) suppressed or absent in the condensed phase, as discussed in Section 6.2, resulting in an increase in some cross sections to conserve the Thomas–Reiche–Kuhn sum rule (Zheng et al. 2016). However, quantifying cross sections in the condensed phase remain an experimental challenge and theoretical calculations are in their infancy.

The study of the role of electron scattering in astrophysics, astrochemistry and astrobiology is likely to expand in the next decades as it provides a challenging but intriguing subject of research.

6.6 ELECTRONS AND NANOTECHNOLOGY

The nanoscale revolution was one of the major advances in science and technology during the late twentieth century. Our ability to manipulate systems down to nanoscales, build and exploit nanoscale particles and structures is at the forefront of current scientific research. Perhaps the most dramatic example of nanoscaling is the fabrication of electronic devices with nanoscale characteristics. The steady shrinking of the size of transistor from microns to 10s of nanometres has been central to the increase in computational efficiency and, the operation of the so-called Moore's law stating that the number of transistors on a microprocessor chip will double every two years or so, has been core to the development of the semiconductor industry. However Moore's law may be coming to an end (Waldrop 2016), requiring new methods for fabricating structures below 10nm, e.g., Focussed Electron Beam Induced Deposition (FEBID) and Extreme Ultraviolet Lithography (EUV). Nevertheless future production of many of the 'chips' required by the semiconductor industry will continue to rely on plasma processing methods in which electron processes are dominant.

6.6.1 Plasma Etching

Fluorocarbon plasmas have dominated semiconductor plasma processing for more than three decades. Much of the technological progress in modern electronics and computational power has been due to advances in our understanding and exploitation of fluorocarbon plasmas. Indeed, the ability of industry to match 'Moore's law' has been due to these improvements. Today some of the most detailed and realistic simulations of any technological process involve the simulation of plasma etching; building microstructures using fluorocarbon plasmas. Such models have been entitled 'virtual factories' (Figure 6.14) since they are sufficiently robust to allow for testing new methodologies for chip manufacture before expensive technical plants are constructed. Such virtual factories may test dozens of different scenarios providing high cost benefit to the industry. This has become particularly important when the choice of feed gas is increasingly restricted due to environmental concerns. A crucial input in these virtual factories is electron scattering data.

Typical fluorocarbons include SF_6, CF_4, C_2F_4 and variants thereof, and accordingly the electron interactions with such species have been extensively studied (Chapter 5) and a detailed but still not complete database of such interactions has been compiled by Makabe and Petrovic (2014). It should be noted that all halogenated species have strong cross sections for electron attachment and hence fluorocarbon plasmas are 'electronegative plasmas' with the anions playing a strong role in subsequent plasma chemistry. Indeed, since DEA is a resonance driven process, changing the electron temperature may alter the relevant DEA pathways allowing the chemistry in the plasma to be 'tuned'.

Unfortunately all of the commonly used fluorocarbon feed gases have high 'Global Warming Potentials' (GWPs). Indeed, SF_6 has the highest GWP (= 23,000) of any known compound, i.e., one molecule of SF_6 is equivalent to 23,000 CO_2 molecules. Accordingly, with increasing awareness of the contribution of such gases to global warming, protocols have been put in place to ban the use of such compounds within the decade. Such protocols provided the semiconductor industry with its greatest challenge since they attack the very basis of the industry. Major research programs have been established to explore alternatives but, to date, the alternatives (e.g., CF_3I, NF_3) are not as efficient as those gases they replace, or may have other drawbacks such as toxicity or reactivity. An alternative is to use non-fluorocarbon species such as NH_3 but the reproducibility and stability of the microstructures such plasmas create remains uncertain, in part due to the lack of a comprehensive database to develop a 'virtual factory' using such feed gases against which to test fabrication conditions.

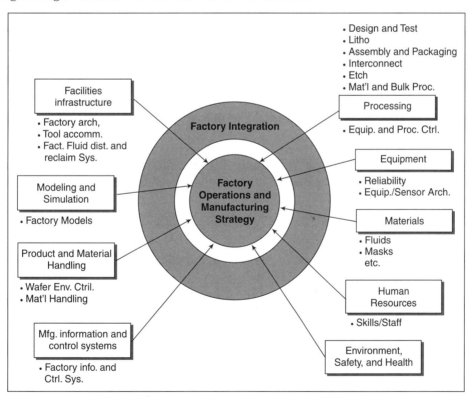

Figure 6.14 A 'virtual factory' is a computerized simulation of a fabrication facility; reproduced from SIA Semiconductor Industry Association Roadmap (1994)

In all such plasmas a core electron induced process is the formation of the etching radicals (CF_x, SF_x or NH_x where x = 1 to 3) hence the characteristics of electron induced dissociation processes forming such radicals must be studied in depth. Furthermore, subsequent to their formation, such radicals may experience electron collisions including excitation, ionization and DEA. However, to date, experiments on such species are limited to only a few discrete cases since it is difficult to produce adequate concentrations of such radicals that are sufficient for current experimental techniques. Accordingly, most of our knowledge of the cross sections for such processes is through theoretical evaluation of the electron scattering cross sections (of which several examples have been given in the previous chapters). At low energies, theoretical studies of such species are challenged both by the ability to create accurate representations of such scattering targets, and the complex and computationally expensive models needed to study electron induced dissociative processes. Thus, despite the great progress in electron scattering theory and creation of 'virtual factories' a full description of electron induced processes and subsequent chemistry in fluorocarbon plasmas remains to be completed.

6.6.2 Next Generation Nanotechnologies

As discussed above, the fabrication of smaller and smaller structures has been the goal of the electronics industry for more than three decades and, since the smaller the structures the stronger the operational power of the manufactured 'chip', Moore's law has ruled the semiconductor industry through this period. However, as the size of the structures falls below 30 nano-metres (nm), traditional manufacturing methods using plasma etching are struggling to meet Moore's law. Hence the urgent need to develop new nano-fabrication methods.

FEBID (Figure 6.15a), and EUVL (Figure 6.15b) as detailed by Mackus et al. (2014) are two complimentary methods that have the potential to achieve resolution better than 10 nm.

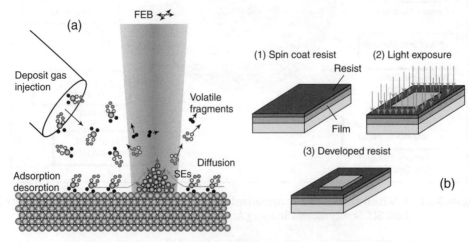

Figure 6.15 Principles of (a) Focused Electron Beam Induced Deposition (FEBID) 1 and (b) Extreme Ultraviolet Lithography (EUVL); from Mackus et al. (2014)

FEBID is a one-step process capable of fabricating 3D nanostructures on uneven surfaces. EUVL, on the other hand, is a multi-step mask based technique suitable for high throughput in conditions of current processing flows.

Both FEBID and EUVL approaches are ideally suited to the fabrication of nanoscale devices and can easily replace and/or can be incorporated in existing fabrication processes, thus providing commercially viable technologies for the most flexible synthesis, fabrication and rapid prototyping of low dimensional materials and structures.

In FEBID, a tightly focused high-energy electron beam impinges onto a surface exposed to a continuous stream of precursor molecules. Electron–molecule interactions decompose these molecules to form a deposit. Ideally a chemically and structurally well-defined (e.g. pure metal) deposit is left behind while the volatile fragments are pumped away. To date, FEBID has relied on precursor molecules developed for chemical vapour deposition (CVD); a process that is governed by thermal decomposition. However, the dissociation mechanisms in FEBID are non-thermal electron-induced reactions and consequently the precursors used for CVD are not optimized for FEBID, leading to the production of nanostructures with undesirable chemical composition, as pointed out by Utke et al. (2008). In FEBID the energy of the primary electron (PE) beam is in the 10s of keV range but the bulk of the chemical reactions are believed to be initiated by low energy secondary electrons (LESEs). These LESEs are created in large numbers when a high-energy electron beam impinges on a substrate and are emitted from the substrate and the deposits surface, through inelastic scattering of the PEs. Indeed, thousands of secondary electrons may be generated by a FEBID, TEM source. This makes the electron induced chemistry that governs FEBID substantially more complicated and also creates a low-energy electron flux outside the focal point of the PE beam (formed through scattered PEs). This, in turn, can cause significant broadening of the structures, beyond the focal width of the electron beam, reducing the ultimate spatial resolution of the respective nano structures. The energy distribution of the LESEs may reach well above 20 eV where dissociative ionization (DI) has high cross sections, but many of the secondary electrons have energies below 10 eV, i.e., below the ionization threshold where electrons fragment molecules by dissociative electron attachment (DEA) and neutral dissociation through electronic excitation (ND).

In EUV processing, the surface to be patterned is initially coated with a viscous liquid solution of a photoresist, creating a uniform, very flat, layer of sub- to few-micrometer thickness. The solvent is then removed by prebaking the coated substrate at fairly low temperatures (usually about 100°C). The dry, uniform photoresist is then exposed to EUV light in a predefined arrangement reflecting the pattern to be produced. EUV exposure causes chemical reactions in the resist such that, after the exposure, either the exposed layer (positive resist) or unexposed layer (negative resist) can be dissolved from the substrate leaving the desired pattern on the substrate. Before the resist is removed with an appropriate solvent (developer) a post-exposure bake (PEB) is conducted. In this step much of the chemistry initiated by the EUV exposure takes place, the exact details of which depend on the chemical structure and composition of the resist. This is especially true for Chemically Amplified Resists in which the initial photo ionization process leads to the formation of 'photo acids' (free protons and protonated species), which diffuse within the resist, limiting the achievable resolution of the nanostructure. Typically, after the PEB the substrate is hard baked to strengthen the remaining photoresist, before final functionalization of the substrate by for instance ion implantation, wet chemical etching, or plasma etching. In EUV, similar to FEBID, a high number of photoelectrons are produced within the resist, and these generate an even larger number of secondary electrons that are pivotal to the chemistry induced in the resist.

In contrast to FEBID, however, where the scattering length of the high energy PEs defines the spatial distribution of the secondary electrons, the initially generated photoelectrons in EUVL already have fairly low energy (less than and around 80 eV). Their spatial distribution is thus defined by their mean free path, which in turn is defined by the chemical composition of the resist and the initial energy of the photoelectrons. To date, the key photoresist targets with resolution, sensitivity and line width roughness, abbreviated as RLS are not simultaneously reached with the traditional chemically amplified photoresists. Innovative conceptualization and deeper fundamental studies are therefore required in order to unlock the full potential of EUVL technology. As electrons play a pivotal role in the chemistry governing the performance of the EUV resists, it is desirable to tune the chemical composition of the resist to i) limit the spatial distribution of the electrons, and, ii) more importantly, to control the chemistry induced by these electrons. Both these goals can be achieved by chemically tuning the susceptibility of the resist material to incident electrons and more specifically by controlling the relative extent of DEA, electron impact excitation, and dissociation and dissociative ionization. The large range of potential chemical species for FEBID and EUV lithography requires large databases for electron interactions with each of these species and, since much of this data needs to be generated theoretically, this becomes a challenge for the electron scattering community.

6.7 SCATTERING UNDER EXTERNAL PLASMA CONFINEMENTS

The largest plasma technology project in human history is the International Thermonuclear Experimental Reactor (ITER) project aimed at establishing fusion as a viable future energy source. The ITER tokamak will measure 24 metres high and 30 metres wide. It will be smaller than a conventional power station but is expected to produce up to 500 MW of thermal power in toroidal fusion plasma with a volume of 800 m^3. The ITER experiment is developed using the concept of the tokamak which was a device first conceived in Moscow, back in the 1960s. A tokamak is a torus or 'doughnut-shaped' continuous tube surrounded by coils that produce a magnetic cage to confine the high-energy plasma. In ITER all the magnetic coils are superconducting and it is expected to bring together all necessary technology for a future fusion power station. ITER is 30 times more powerful than the previous largest tokamak – the Joint European Torus (JET) – and will produce ten times more power than is necessary to maintain plasma at fusion temperatures in order to demonstrate the physics of burning plasma: plasma that is heated by internal fusion reactions rather than external heating. Engineers will then be able to test the heating, control, diagnostic and remote maintenance systems that will be needed for a real fusion based power station.

Atomic collision processes play a large role in both magnetic and inertial confinement fusion research. There are four principal areas of fusion research where atomic processes are important: (1) the hot central plasma where the fusion reactions are expected to occur, (2) the plasma edge where the plasma interacts with the external environment, (3) plasma heating methods used to produce the hot plasma, and (4) diagnostic techniques used to measure the physical properties of the plasma. In all cases, electron collisions play an important role and, of particular relevance, are electron interactions with hydrogen (atomic and molecular) and its isotopes deuterium and tritium. The high temperatures within the plasma are such that molecular hydrogen is present in highly vibrationally excited states (defined by vibrational quantum number υ) and hence electron scattering cross sections (and rate constants) for electron scattering from ro-vibrationally excited $H_2(\upsilon)$ are required. Such cross sections may be significantly different from those of the ground state, indeed DEA leading to formation of $H + H^-$ may change by orders of magnitude while there are also strong isotope effects with the formation of anions via the lowest (4eV) resonance in D_2 being 1/100 of that of corresponding H_2, a process that is important in developing neutral beam injectors. Experiments using radioactive tritium are, of course, sparse and so current data sets of electron interactions with tritium are mainly theoretical. Furthermore even for as simple a molecular system as H_2, calculations of cross sections provide severe challenges (e.g., the cross section for excitation and subsequent dissociation of the lowest excited state resulting in two ground state atoms). Due to the high degree of ionization in the fusion plasma, dissociative recombination (DR) is an important process but cross sections (particularly for ro-vibrationally excited cations) remain a subject of debate; as do the fluorescent yields of H atoms excited in the plasma, so important in calibrating several spectroscopic diagnostic techniques.

At the plasma edge, the hydrocarbon species become important as do beryllium hydrides (beryllium being the material chosen to line the tokamak wall facing the plasma). Electron interactions with many of these species (particularly CH_x radicals and beryllium hydrides) are largely unquantified and often rely on theoretical calculations that are poorly benchmarked. Semi-empirical calculations (Vinodkumar et al. 2006, Kim and Rudd 1994) of electron impact ionization cross sections appear, in general, to be reliable though they do not give branching ratios. Recently, the available codes, such as quantemol and the Independent Atom approximation, have been shown to give good estimates of elastic and summed inelastic cross sections over a wide energy range but excitation cross sections (ro-vibrational and electronic) and dissociative pathways (DEA, etc.,) remain largely uncharacterized and constitute a research area that has to be explored concurrently with the development of the ITER facility.

At this stage, consideration has to be given to a slightly different problem of electron scattering from neutral atoms or molecules embedded in the external plasma environment. Plasma curtails long range potentials, and if strong enough, influences the structural properties of the embedded atomic/molecular species. As a special case, consider a weak Debye plasma, in which the collective effect of the surrounding on an embedded electric charge is characterized by the Debye screening length Λ_D, such that its Coulomb potential is screened off by a factor $\exp(-r/\Lambda_D)$, at distance r. Elastic scattering of electrons by H atoms and separately by H_2 molecules embedded in weak plasma was studied by Modi et al. (2015). In a high energy approximation, the effect of external weak plasma was examined by these authors, by considering selected values of Debye length Λ_D from 5 to $20a_0$. At $\Lambda_D = 5\ a_0$ the decrease in the forward elastic differential cross sections in both the targets was found to be about 20%, while the elastic TCS decreased by about 10%.

6.8 BIOMOLECULAR TARGETS AND RADIATION DAMAGE

One of the most exciting and intriguing discoveries in atomic and molecular physics in the twenty-first century has been the finding that radiation induced damage can be postulated in terms of discrete collision phenomena, thus enabling conceptualization of nanoscale or molecular dosimetry and setting the foundations for future radiotherapies.

The basis of these studies is the pioneering work of Sanche and co-workers who reported electron induced damage on strands of DNA (Sanche 2005). The DNA helix can be ruptured in two ways: one leading to single strand breaks (or SSBs); the other producing double strand breaks (or DSBs), as shown in Figure 6.16. Boudaïffa et al. (2000) showed that contrary to expectation, low energy (< 10 eV) electrons could produce both SSBs and DSBs and within these damage curves, structures were revealed that are reminiscent of 'resonances' seen in electron–molecule scattering (Figure 6.13).

These results challenged many assumptions of the radiation chemistry community and our understanding of how DNA, the most important molecule in the cell nucleus, is damaged when irradiated by ionizing radiation. Instead of the primary radiation (X-ray, gamma ray, UV light) causing the majority of the damage; the secondary electrons produced by such radiation, may induce 'direct' damage while also producing the free radicals that produce 'indirect damage', as shown in Figure 6.17. In a new nanoscale model of radiation damage, the primary form of ionising radiation (X-ray, UV photon, ion) is incident on the cell, and may strike a DNA molecule leading to 'direct damage' or may pass through, creating secondary species. Such species include secondary electrons or chemically reactive species such as OH and H_2O_2 liberated from the water in the cell and the damage they induce is known as 'indirect' damage. Although it is subject to some debate and open to further study, roughly 1/3 of the nascent damage is commonly attributed to direct and 2/3 to indirect damage. Many therapeutic studies therefore seek to enhance free radical liberation in the tumour to cause its death.

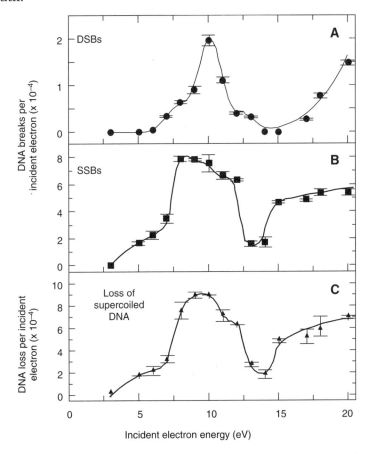

Figure 6.16 Electron induced damage in DNA showing production of Single Strand Breaks (SSB), Double Strand Breaks (DSB), and loss of supercoiled DNA for electron energies between 3 and 20 eV spanning the ionization energy of DNA constituent molecules (Boudaïffa et al. 2000)

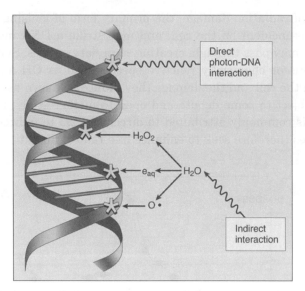

Figure 6.17 Schematic diagram showing direct and indirect radiation damage

6.8.1 Radiotherapy and Treatment Planning

In radiotherapy, common sources include X-rays, gamma rays, fast electrons, ion beams such as 60 to 400 MeV, protons, and C^+ ions (Tsujii 2017). Conventionally, the energy of diagnostic and therapeutic gamma- and X-rays is expressed in kilovolts or megavolts (kV or MV), whilst the energy of therapeutic particles is expressed in terms of mega-electronvolts (MeV). In the first case, this voltage is the maximum electric potential used by a linear accelerator to produce the photon beam. The beam is made up of a spectrum of energies: the maximum energy is approximately equal to the beam's maximum electric potential times the electron charge. Thus a 1 MV beam will produce photons of no more than about 1 MeV and the mean X-ray energy is only about 1/3 of the maximum energy. Megavoltage X-rays (1 to 25 MeV) are by far the most common in radiotherapy for treatment of a wide range of cancers. Superficial (35 to 60 keV) and ortho-voltage (200 to 500 keV) X-rays find application for the treatment of cancers, at or close to the skin surface, because the penetration depths of these X-rays are lower than those for megavoltage X-rays.

The energy of the incident radiation is, however, not the major difference between photon and particle therapy; rather it is how the energy is deposited in the body. Photon therapies see a gradual decrease in the energy deposited as a function of depth whereas particle (or hadron) therapy sees the maximum energy deposited in a region at depth, the so-called Bragg Peak, as depicted in Figure 6.18. Thus hadron therapies have the potential to deposit the energy within the tumour inside the body allowing greater radiation doses to be applied to the tumour itself, while minimizing irradiation to adjacent normal tissues. In the region beyond the end of the peak, almost no dose is deposited with protons, while a small dose is deposited with carbon ions. This is because primary carbon ions undergo nuclear interactions and fragment into particles with a lower atomic number, producing a fragmentation tail beyond the peak (Tsujii 2017). Hadron therapy centres are increasing in number around the world with over 100,000 patients by 2016, but these are small numbers when compared to the several million receiving photon therapy.

All radiotherapy treatments are strongly regulated and treatment plans arc sct using 'Radiation Therapy Treatment Planning Systems' which calculate the required monitor units to deliver a prescribed dose to a specific area, and the distribution of irradiation this will create in the body. These computer codes are based on a chemical and physical knowledge of how radiation interacts with tissue. At present these tools are mostly based on macro-dosimetry but hadron therapy requires a more molecular and nano-dosimetry approach. Accordingly, there is a need to understand the interaction of discrete particles with DNA and its constituent molecules (Brandas and Sabin 2006).

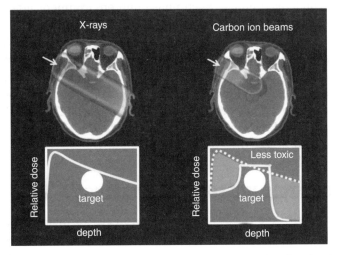

Figure 6.18 Schematic showing the dose deposition by X-rays and Carbon-ion beams; note the concentration of dose deposition in the so-called Bragg Peak by carbon ions

Often, radiotherapy can be improved by use of radio-sensitizers, which are molecules or other material (e.g. a nanoparticle) that make tumour cells more sensitive to radiation therapy allowing greater effective damage for the same (or smaller) dose(s) of radiation, reducing harmful effects on normal cells and reducing the side effects of radiation. Many of the most effective radio-sensitizers contain halogen atoms and have larger DEA cross sections providing evidence for the hypothesis of Sanche and co-workers that electrons do play a major role in inducing DNA damage. This is also supported by the evidence for nanoparticles (NPs) being good radio-sensitizers with the incident irradiation liberating secondary electrons from the NPs (Haume et al. 2016).

In any X-ray or hadron passage through tissue many tens or even hundreds of thousands of electrons will be liberated with energies from >1 keV to 'thermal', with low energy electrons becoming 'solvated electrons'. Accordingly, there has been an explosion in the study of electrons with biomolecules, with DEA processes being particularly prevalent and (dissociative) ionization and elastic scattering data being recorded by many groups. There is less data on electron impact excitation of biomolecules and little on neutral dissociation and dipolar dissociation/ion pair formation. Theoretical models of electron biomolecule interactions have made great progress in the past decade (Baccarelli et al. 2011) in particular in identifying resonances at low energy. However it is important to recognize that these experiments and calculations have largely assumed that the target is in a gaseous state whereas in the cellular environment the target molecule is influenced by its surrounding medium (as explained in Section 6.2, above). Experiments of biomolecules in clusters, including water clusters, provide a step towards exploring the effect of hydration in the cell as do experiments on these molecules in the condensed phase. Such phase and environment effects are shown to strongly perturb electron processes, particularly DEA. Theoretical methods using the independent atom approximation have provided some data on ionization and elastic scattering cross sections, at least at energies above the ionization threshold. Extending calculations to study more complex biomolecules is a priority and recently the first full R-matrix calculation of a complex system of two hydrated biomolecules (pyridine and thymine) has been reported by Sieradzka and Gorfinkiel (2017). Modelling low energy (0–100 eV) electron and positron tracks in biologically relevant media has been discussed by Blanco et al. (2013).

6.9 POSITRON ATOM/MOLECULE SCATTERING

While this book is dedicated to the study of electron scattering, a complimentary field is the study of the electron anti-particle, the positron. The study of antimatter is an important aspect of astrophysics since one of the major questions arising from the study of the birth and propagation of our universe is why there is so little antimatter remaining when matter and antimatter were formed in the big bang at the same/ similar rates. Furthermore, by studying antimatter it is possible to gain further insights into many of the fundamental electron–matter collisions such as atomic/ molecular excitation where positron collisions are free of 'exchange' effects while the static and polarization force fields are opposite in the case of the positrons in contrast to electrons.

Positrons are a natural product of radioactive decay, for example from ^{22}Na, and may be collimated and/or trapped using same techniques as for electrons, thus allowing positron beams to be prepared for scattering experiments. However the fluence of a positron beam is much less than that of electron beams, since the nascent flux from available positrons sources is low and the emitted positrons have very high energy requiring 'moderators' to slow the positrons to the lower energies needed for scattering experiments. Typical positron beams may have intensities of only a few 1000 per second limiting experiments to total scattering and ionization studies. The development of positrons traps (Danielsson et al. 2015) has allowed larger fluxes of positrons to be built up allowing positron excitation of atoms and molecules to be studied, as well as other forms of antimatter (e.g. antihydrogen) to be created and studied. The resolution of the positron beam is often low compared with electrons (100s of meV) since the flux does not allow monochromatic techniques developed for electrons to be adopted; however by modifying the trapping potential of a trap, resolutions of <100 meV have been achieved allowing vibrational excitation of molecules by the positrons to be observed (Danielson et al. 2015).

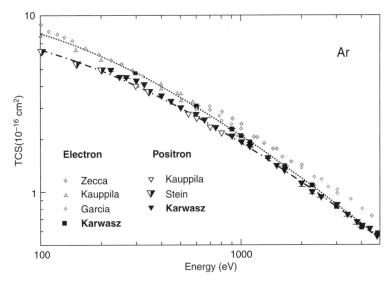

Figure 6.19 Total scattering cross section in Argon using electrons and positrons; the two cross sections merge at high energies (>200 eV) (Karwasz 2002)

At some higher energy where the positrons are moving rapidly past the target, such that there is little time for the difference in sign of the charge to be significant, the total scattering cross sections for electrons and positrons are expected to be the same. For most targets this is found to hold for energies above 200 eV, as shown in Figure 6.19 vide Karwasz et al. (2002). Similarly, ionization cross sections at high energies are the same for energies >200–300 eV.

At low energies the different sign of the charge becomes crucially important. As mentioned above, charge exchange is absent in positron excitation of atoms and molecules, and so, positrons cannot excite triplet states of helium. Low energy electron scattering is dominated by the formation of resonances in which the incident electron is captured by the target to form a transient negative ion. There is no comprehensive experimental evidence for positron induced resonances but positrons are readily captured by electrons to form the quasi-atom positronium (Ps), an unstable electron-positron bound system whose excited states have energies half of that of atomic hydrogen. All positron scattering cross sections show structures at the threshold for positronium formation. Although positronium is short lived (140 nano-seconds), it is possible to form positronium beams which may be used to scatter particles from atoms and molecules (Laricchia et al. 1987, Fayer et al. 2015). Ps scattering often shows similarities with electron scattering from the same target (Shipman et al. 2017). At very low energies (<1eV) positrons annihilate with electrons in the target releasing high energy gamma rays. The cross sections for the annihilation processes may be high, e.g. for hydrocarbons, Gribakin et al. (2010), a mechanism that is considered similar to the resonance scattering processes observed at close to zero energy in DEA in electron scattering.

Theoretical calculations of positron atom/molecule collisions have proven to be more difficult than initially expected since at low energies it is not a simple case of just changing the charge of the projectile. For positronium formation, annihilation rates, the absence of exchange and the changes in induced polarization require the scattering problem to be reformatted if low energy scattering cross sections are to be quantified. At high energies, where electron and positron scattering cross sections are expected to merge, the semi-empirical methods described in this book to generate electron total and ionization cross sections can be used to estimate analogous positron cross sections.

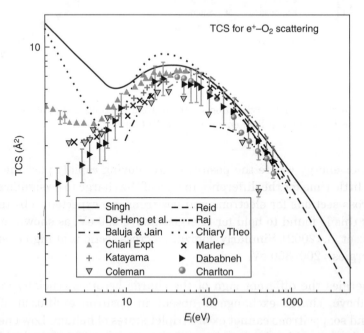

Figure 6.20   Positron scattering from molecular oxygen showing data of Singh and Antony (2017) using a modified spherical complex optical potential formalism, compared to a range of data (see Singh and Antony 2017)

Recently Antony and co-workers have used a modified version of spherical complex optical potential formalism to calculate total positron scattering cross sections over a wide energy range from near positronium formation threshold to 5000 eV for rare gas atoms and several molecules such as C_2N_2, O_2, CH_4, CO, CO_2, H_2, N_2O and NO (Singh et al. 2016, Singh and Antony 2017). Further work has extended these calculations to explore inelastic channels such as positronium (Ps) formation, direct ionization, and total ionization. Singh and Antony (2017, and also references therein) found good agreement up to energies as low as 10 eV, as in Figure 6.20.

Earlier Kothari and Joshipura (2010, 2011, 2012) formulated the CSP-ic, for positron (e^+) scattering from inert gas atoms Ne, Ar, and Kr, and the isoelectronic molecules N_2 and CO. Appropriate models for polarization and absorption were incorporated and the total inelastic cross section was defined to be $Q_{inel} = Q_{ion} + Q_{Ps} + \Sigma Q_{exc}$, where Q_{Ps} is the positronium formation cross section, known for a few common atoms and molecules. Results from Kothari and Joshipura (2010) are reproduced for neon in Figure 6.21, wherein comparisons with theoretical and experimental data have been made.

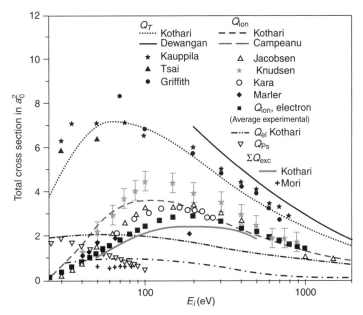

Figure 6.21 Various total cross sections of positron–neon scattering (Kothari and Joshipura 2010); Q_T: *dotted curve*: CSP-ic; *continuous*: Dewangan and Walters (1977); measured Q_T: *stars*: Kauppila et al. (1981); *triangles*: Tsai (1976); *bullets*: Griffith et al. (1979); Lower data-sets Q_{ion}: *dashed curve*: CSP-ic; *faint curve*: Campeanu et al. (2001); *triangles*: Jacobsen et al. (1995); *faint stars*: Knudsen et al. (1990); *circles*: Kara et al. (1997); *diamonds*: Marler et al. (2005); *squares*: average experimental Q_{ion} for electrons, Krishnakumar and Srivastava (1988); Q_{el}: *dash-dots*: CSP-ic; Q_{Ps}: *inverted triangle*: Marler et al. (2005); ΣQ_{exc}: *dash-dots*: CSP-ic, + Q_{exc} by Mori and Sueoka (1994)

The low energy behaviour of Q_T and Q_{el} for positrons is remarkably different from that for electrons. The CSP-ic Q_{ion} for positrons (Figure 6.21) are in general accord with experimental data and are higher than the electron Q_{ion} up to about 300 eV. For 100 eV positrons incident on neon atoms, the cross sections Q_{el}, Q_{ion}, ΣQ_{exc} and Q_{Ps} are found to be 29, 52, 13 and 6% of Q_T respectively (Kothari and Joshipura 2010).

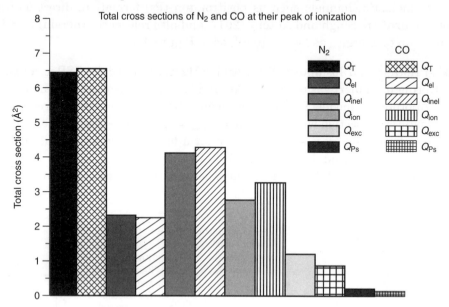

Figure 6.22 A composite bar-chart showing all the TCS of positron scattering with N_2 and CO at 100 eV (Kothari and Joshipura 2011)

Positron scattering and ionization of N_2 and CO were examined by Kothari and Joshipura (2011) in CSP-ic and their results were found to be generally consistent with experimental and other data. An interesting comparison of all the 100 eV TCS of these two isoelectronic molecules is exhibited in the composite bar-chart, Figure 6.22, reproduced from Kothari and Joshipura. The cross sections Q_T, Q_{el}, Q_{inel} and Q_{Ps} are seen to be quite similar in this isoelectronic pair, but the Q_{ion} of CO are a little larger than that of N_2, with a corresponding difference in ΣQ_{exc}.

Positrons are used as an analytical tool using positron annihilation spectroscopy (PAS) in materials research to detect variations in density, defects, displacements, or even voids, within a solid material. Positrons are basic in the use of positron emission tomography (PET) scanners that are used in hospitals to create detailed three-dimensional images of metabolic activity within the human body by detecting the gamma rays, emitted indirectly by a positron-emitting radionuclide (tracer) placed into the patient by ingestion. Indeed most positron sources are produced for use in the medical community. Future experimental research will therefore include positron studies with the larger biomolecules currently being studied by electrons (Section 6.8 above) and, as sources and trap technologies develop, may be extended to cluster targets. Positron induced chemistry is in its infancy but as positron induced fragmentation studies develop, it may be expected to grow in a manner similar to the advances in electron induced chemistry (Jean et al. 2003).

6.10 CONCLUSIONS: THE FUTURE OF ELECTRON SCATTERING

Electrons are omnipresent in nature and electron induced processes drive many of the most important processes of life, such as photosynthesis (Roach and Krieger-Liszkay 2014). In this concluding chapter we have presented some of the varied applications of electron scattering, and the electron collision cross sections that are required to model and understand such processes, and have mentioned specific data needs where relevant. There are many other situations where knowledge of electron scattering processes is required. These include the design of lighting; the development of plasma sources for surgery and dentistry, covered by Fridman and Friedman (2013); the design of hypersonic space craft capable of re-entry, as discussed by Celiberto et al. (2014); and in combustion processes, where the removal of electrons by electronegative gases provides a means to extinguish a fire without withdrawing oxygen and is thus used in aircraft to extinguish cabin fires.

The study of electron scattering from atoms and molecules has grown rapidly since the Frank and Hertz experiment in 1914 and the first quantum mechanical treatments in the 1930s. The wide range of applications ensure that the study of such processes will be important in times to come. Much of the data required for such applications will necessarily be generated by theoretical methods, as explained in this book, with benchmarks being provided by a few definitive experiments.

In conclusion, we look back and wonder at the long conceptual journey from hydrogen to complex atoms, from the small and the common to large and exotic molecular systems, which include fascinating radicals and metastable species. Larger targets such as biomolecules offer challenges as well as opportunities. Collisions and scattering that we dealt with in the first five chapters are essentially micro-level phenomena but micro rules the macro, and that was the guiding principle behind the discourses of this concluding chapter. For, it is the basic knowledge that opens up possibilities of potential application in diverse fields of technology today.

Though we are in the second decade of the twenty-first century, it would be most appropriate to end with a famous and relevant quote from none other than Richard Feynman.

'We are at the very beginning of time for the human race. It is not unreasonable that we grapple with problems. But there are tens of thousands of years in the future. Our responsibility is to do what we can, learn what we can, improve the solutions, and pass them on.'

– Richard P. Feynman

✪ ✪ ✪

BIBLIOGRAPHY

Abdel-Naby, S. A., M. S. Pindzola, A. J. Pearce, C. P. Ballance, and S. D. Loch. 2014. 'Electron-impact ionization of the N atom.' *Journal of Physics B: Atomic, Molecular and Optical Physics* 48 (2): 025203.

Adamczyk, B., A. J. H. Boerboom, B. L. Schram, and J. Kistemaker. 1966 'Partial ionization cross sections of He, Ne, H_2, and CH_4 for electrons from 20 to 500 eV.' *The Journal of Chemical Physics* 44 (12): 4640–4642.

Adamovich I., S.D. Baalrud, A. Bogaerts, P.J. Bruggeman, M. Cappelli, V. Colombo, U. Czarnetzki, et al. 2017. *Journal of Physics D: Applied Physics* 50 (32): 323001

Ali, M. A., Y-K. Kim, W. Hwang, N. M. Weinberger, and M. Eugene Rudd. 1997. 'Electron-impact total ionization cross sections of silicon and germanium hydrides.' *The Journal of Chemical Physics* 106 (23): 9602–9608.

Ali, M. A., and Yong-Ki Kim. 2008. 'Ionization cross sections by electron impact on halogen atoms, diatomic halogen and hydrogen halide molecules.' *Journal of Physics B: Atomic, Molecular and Optical Physics* 41 (14): 145202.

Allan, Michael. 2007. 'Absolute angle-differential elastic and vibrational excitation cross sections for electron collisions with tetrahydrofuran.' *Journal of Physics B: Atomic, Molecular and Optical Physics* 40 (17): 3531.

———. 2010. 'Electron collisions with CO: elastic and vibrational excitation cross sections.' *Physical Review A* 81 (4): 042706.

Almeida, D. P., M. A. Scopel, R. R. Silva, and A. C. Fontes. 2001. 'Single and double ionisation of mercury by electron impact.' *Chemical Physics Letters* 341 (5): 490–494.

Antony, B. K., K. N. Joshipura, N. J. Mason, and Jonathan Tennyson. 2004. 'R-matrix calculation of low-energy electron collisions with LiH.' *Journal of Physics B: Atomic, Molecular and Optical Physics* 37 (8): 1689.

Antony, B. K., K. N. Joshipura, and N. J. Mason. 2004. 'Electron impact ionization studies with aeronomic molecules.' *International Journal of Mass Spectrometry* 233 (1): 207–214.

———. 2005. 'Total and ionization cross sections of electron scattering by fluorocarbons.' *Journal of Physics B: Atomic, Molecular and Optical Physics* 38 (3): 189.

Anzai, K., H. Kato, Masamitsu Hoshino, H. Tanaka, Y. Itikawa, L. Campbell, Michael J. Brunger, et al. 2012. 'Cross section data sets for electron collisions with H_2, O_2, CO, CO_2, N_2O and H_2O.' *The European Physical Journal D-Atomic, Molecular, Optical and Plasma Physics* 66 (2): 1–8.

Ariyasinghe, W. M., and D. Powers. 2002. 'Total electron scattering cross sections of CH_4, C_2H_2, C_2H_4, and C_2H_6 in the energy range 200–1400 eV.' *Physical Review A* 66(5): 052716.

Armentrout, Peter B., Susan M. Tarr, Ariel Dori, and Robert S. Freund. 1981. 'Electron impact ionization cross section of metastable N_2 (A Σu^+).' *The Journal of Chemical Physics* 75 (6): 2786–2794.

Ashrafi, M., B. S. Lanchester, D. Lummerzheim, N. Ivchenko, and O. Jokiaho. 2009. 'Modelling of $N_2$1P emission rates in aurora using various cross sections for excitation.' *Annales Geophysicae: Atmospheres, Hydrospheres and Space Sciences* 27 (6): 2545.

A'Hearn, M.F., S. Hoban, P.V. Birch, C. Bowers, R. Martin, and D.A. Klinglesmith. 1986.'Cyanogen jets in comet Halley.' *Nature* 324 (6098): 649–651.

Baccarelli I., Ilko Bald, I. Gianturco, E. Illenberger, and J. Kopyra. 2011. 'Electron-induced damage of DNA and its components: Experiments and theoretical models.' *Physics Reports* 508 (1): 1–44

Baek, W. Y., M. Bug, H. Rabus, E. Gargioni, and B. Grosswendt. 2012. 'Differential elastic and total electron scattering cross sections of tetrahydrofuran.' *Physical Review A* 86 (3): 032702.

Baek, W. Y., and B. Grosswendt. 2003. 'Total electron scattering cross sections of He, Ne and Ar, in the energy range 4 eV–2 keV.' *Journal of Physics B: Atomic, Molecular and Optical Physics* 36 (4): 731.

Bagenal, F., M. Horányi, D. J. McComas, R. L. McNutt, H. A. Elliott, M. E. Hill, L. E. Brown et al. 2016. 'Pluto's interaction with its space environment: Solar wind, energetic particles, and dust.' *Science* 351 (6279): aad9045.

Baiocchi, Frank A., Robert C. Wetzel, and Robert S. Freund. 1984. 'Electron-Impact Ionization and Dissociative Ionization of the CD_3 and CD_2 Free Radicals.' *Physical Review Letters* 53 (8): 771.

Baluja, K. L., A. Jain, V. Di Martino, and F. A. Gianturco. 1992. 'Total (elastic plus inelastic) cross sections for electron scattering with CF_4 and GeH_4 molecules at (10 – 5000) eV.' *EPL (Europhysics Letters)* 17 (2): 139.

Bart, Mark, Peter W. Harland, James E. Hudson, and Claire Vallance. 2001. 'Absolute total electron impact ionization cross sections for perfluorinated hydrocarbons and small halocarbons.' *Physical Chemistry Chemical Physics* 3 (5): 800–806.

Bartlett, Philip L., and Andris T. Stelbovics. 2002. 'Calculation of electron-impact total-ionization cross sections.' *Physical Review A* 66 (1): 012707.

———. 2004. 'Electron-impact ionization cross sections for elements Z=1 to Z=54'. *Atomic Data and Nuclear Data Tables* 86 (2): 235–265.

Basner, R., R. Foest, M. Schmidt, F. Sigeneger, P. Kurunczi, K. Becker, and H. Deutsch. 1996. 'Electron impact ionization of tetramethylsilane (TMS).' *International Journal of Mass Spectrometry and Ion Processes* 153 (1): 65–78.

Basner, R., M. Gutkin, J. Mahoney, V. Tarnovsky, H. Deutsch, and K. Becker. 2005. 'Electron-impact ionization of silicon tetrachloride ($SiCl_4$).' *The Journal of Chemical Physics*, 123 (5): 054313.

Basner, R., M. Schmidt, E. Denisov, K. Becker, and H. Deutsch. 2001. 'Absolute total and partial cross sections for the electron impact ionization of tetrafluorosilane (SiF_4).' *The Journal of Chemical Physics* 114 (3): 1170–1177.

Basner, R., M. Schmidt, H. Deutsch, V. Tarnovsky, A. Levin, and Kurt Becker. 1995. 'Electron impact ionization of the SO_2 molecule.' *The Journal of Chemical Physics* 103 (1): 211–218.

Basner. R., M. Schmidt, V. Tarnovsky, and K. Becker. 1997. 'Dissociative ionization of silane by electron impact.' *International Journal of Mass Spectrometry and Ion Processes* 171 (1-3): 83–93.

Belic, D. S., and M. V. Kurepa. 1985. 'Electron–Hydrogen Sulfide Total Ionization and Electron Attachment Cross Sections'. *Fizika* 17: 117.

Benda, Jakub, and Karel Houfek. 2014. 'Collisions of electrons with hydrogen atoms I. Package outline and high energy code.' *Computer Physics Communications* 185 (11): 2893–2902.

Beran, J. A., and L. Kevan. 1969. 'Molecular electron ionization cross sections at 70 eV.' *The Journal of Physical Chemistry* 73 (11): 3866–3876.

Bernhardt, Ph, and H. G. Paretzke. 2003. 'Calculation of electron impact ionization cross sections of DNA using the Deutsch–Märk and Binary–Encounter–Bethe formalisms.' *International Journal of Mass Spectrometry* 223: 599–611.

Bettega, Márcio HF, Adriana R. Lopes, Marco AP Lima, and Luiz G. Ferreira. 2006. 'Electron collisons with cyclobutane.' *Brazilian Journal of Physics* 36 (2B): 570–575.

Bhardwaj, Anil, and Vrinda Mukundan. 2015. 'Monte Carlo model for electron degradation in methane gas.' *Planetary and Space Science* 111: 34–43.

Bhowmik, P. 2012. 'Theoretical studies on electron scattering processes with important atomic and molecular targets.' Ph.D. dissertation, Sardar Patel University Vallabh Vidyanagar India; http://shodhganga.inflibnet.ac.in/handle/10603/7363

Bhoumik, P., R. Dawda, and K. N. Joshipura. 2014. 'Electron scattering with polyatomic molecules: screening corrections in group additivity rule'. In *Electron Collision Processes in Atomic and Molecular Physics*, edited by P. C. Minaxi Vinodkumar, p. 74. New Delhi: Narora Publishing House.

Bhoumik, P., Ramji H. Patel and K. N. Joshipura. 2011. 'Correlations among atomic properties and estimates on exotic atoms'. *Prajna - Journal of Pure and Applied Sciences* 19: 52–55.

Bhutadia, Harshad, Ashok Chaudhari, and Minaxi Vinodkumar. 2015. 'Electron impact ionization cross sections for atomic and molecular allotropes of phosphorous and arsenic.' *Molecular Physics* 113 (23): 3654–3662.

Blaauw, H. J., R. W. Wagenaar, D. H. Barends, and F. J. De Heer. 1980. 'Total cross sections for electron scattering from N_2 and He.' *Journal of Physics B: Atomic and Molecular Physics* 13 (2): 359.

Blaha, M., and J Davis. 1975. 'Elastic scattering of electrons by oxygen and nitrogen at intermediate energies.' *Physical Review* A 12 (6): 2319.

Blair, Shamus A., and Ajit J. Thakkar. 2014. 'Relating polarizability to volume, ionization energy, electronegativity, hardness, moments of momentum, and other molecular properties.' *The Journal of Chemical Physics* 141 (7): 074306.

Blanco, F., and G. García. 1999. 'Improved non-empirical absorption potential for electron scattering at intermediate and high energies: 30–10000 eV.' *Physics Letters A* 255 (3): 147–153.

———. 2002. 'Improvements on the imaginary part of a non-empirical model potential for electron scattering (30 to 10000 eV energies).' *Physics Letters A* 295 (4): 178–184.

———. 2003. 'Screening corrections for calculation of electron scattering from polyatomic molecules.' *Physics Letters A* 317 (5): 458–462.

Blanco, Francisco, Antonio Munoz, Diogo Almeida, Filipe Ferreira da Silva, Paulo Limao-Vieira, Martina C. Fuss, Ana G. Sanz, and Gustavo García. 2013. 'Modelling low energy electron and positron tracks in biologically relevant media.' *The European Physical Journal D* 67 (9): 199.

Bodewits, Dennis, Luisa M. Lara, Michael F. A'hearn, Fiorangela La Forgia, Adeline Gicquel, Gabor Kovacs, Jörg Knollenberg et al. 2016. 'Changes in the physical environment of the inner coma of 67P/Churyumov–Gerasimenko with decreasing heliocentric distance.' *The Astronomical Journal* 152 (5): 130.

Bodo, Enrico, Franco A. Gianturco and Rocco Martinazzo. 2003. 'The gas-phase lithium chemistry in the early universe: elementary processes, interaction forces and quantum dynamics.' *Physics Reports* 384 (3): 85–119.

Bolorizadeh, M. A., C. J. Patton, M. B. Shah, and H. B. Gilbody. 1994. 'Multiple ionization of copper by electron impact.' *Journal of Physics B: Atomic, Molecular and Optical Physics* 27 (1): 175.

Bouchiha D., L. G. Caron, J. D. Gorfinkiel, and L. Sanche. 2008. 'Multiple scattering approach to elastic low-energy electron collisions with the water dimer'. *Journal of Physics B: Atomic Molecular and Optical Physics* 41 (4): 045204.

Boudaïffa, Badia, Pierre Cloutier, Darel Hunting, Michael A. Huels, and Léon Sanche. 2000. 'Resonant formation of DNA strand breaks by low-energy (3 to 20 eV) electrons.' *Science* 287 (5458): 1658–1660.

Boyer, Michael C., Nathalie Rivas, Audrey A. Tran, Clarissa A. Verish, and Christopher R. Arumainayagam. 2016. 'The role of low-energy (≤ 20 eV) electrons in astrochemistry.' *Surface Science* 652: 26–32.

Branchett, Susan E., Jonathan Tennyson, and Lesley A. Morgan. 1990. 'Electronic excitation of molecular hydrogen using the R-matrix method.' *Journal of Physics B: Atomic, Molecular and Optical Physics* 23 (24): 4625.

Bransden, Brian Harold, and Charles Jean Joachain. 2003. *Physics of Atoms and Molecules.* Noida: Pearson Education India.

Brawley, S. J., S. E. Fayer, M. Shipman, and G. Laricchia. 2015. 'Positronium production and scattering below its breakup threshold'. *Physical review letters* 115 (22): 223201.

Bray, Igor, Dmitry V. Fursa, A. S. Kadyrov, A. T. Stelbovics, A. S. Kheifets, and A. M. Mukhamedzhanov. 2012. 'Electron-and photon-impact atomic ionisation.' *Physics Reports* 520 (3): 135–174.

Brescansin, L. M., P. Rawat, I. Iga, M. G. P. Homem, M. T. Lee, and L. E. Machado. 2003. 'Elastic and absorption cross sections for electron scattering by ethylene in the intermediate energy range.' *Journal of Physics B: Atomic, Molecular and Optical Physics* 37 (2): 471.

Bromberg, J. Philip. 1970. 'Absolute Differential Cross Sections of Elastically Scattered Electrons. III CO and N_2 at 500, 400, and 300 eV.' *The Journal of Chemical Physics* 52 (3): 1243–1247.

Brook, E., M. F. A. Harrison, and A. C. H. Smith. 1978. 'Measurements of the electron impact ionisation cross sections of He, C, O and N atoms.' *Journal of Physics B: Atomic and Molecular Physics* 11 (17): 3115.

Brown, Ian G. (ed.). 2004. *The Physics and Technology of Ion Sources.* Weinheim: John Wiley & Sons.

Brunger, Michael James. 2007. 'Comment on 'Electron scattering and ionization of NO, N_2O, NO_2, NO_3 and N_2O_5 molecules: theoretical cross sections'.' *Journal of Physics B: Atomic, Molecular and Optical Physics* 40 (10): 1951.

———. 2017. 'Electron scattering and transport in biofuels, biomolecules and biomass fragments.' *International Reviews in Physical Chemistry* 36 (2): 333–376.

Brunger, Michael J., and Stephen J. Buckman. 2002. 'Electron–molecule scattering cross sections. I. Experimental techniques and data for diatomic molecules.' *Physics Reports* 357 (3): 215–458.

Bull, James N., Peter W. Harland, and Claire Vallance. 2011. 'Absolute total electron impact ionization cross sections for many-atom organic and halocarbon species.' *The Journal of Physical Chemistry A* 116 (1): 767–777.

Bunge, Carlos F., Jose A. Barrientos, and A. Vivier Bunge. 1993. 'Roothaan-Hartree-Fock ground-state atomic wave functions: Slater-type orbital expansions and expectation values for Z= 2-54.' *Atomic Data and Nuclear Data Tables* 53 (1): 113–162.

Burke, Philip George. 2011. *R-Matrix Theory of Atomic Collisions: Application to Atomic, Molecular and Optical Processes.* Heidelberg: Springer Science & Business Media.

Burke, P. G., Philippe Francken, and C. J. Joachain. 1991. 'R-matrix-Floquet theory of multiphoton processes.' *Journal of Physics B: Atomic, Molecular and Optical Physics* 24 (4): 761.

Bussery-Honvault, B., and V. Veyret. 1998. 'Comparative studies of the lowest singlet states of $(O_2)_2$ including ab initio calculations of the four excited states dissociating into O_2 ($^1\Delta_g$) + O_2 ($^1\Delta_g$).' *The Journal of Chemical Physics* 108 (8): 3243–3248.

Byron Jr, F. W., C. J. Joachain, and R. M. Potvliege. 1985. 'Elastic and inelastic scattering of electrons and positrons by atomic hydrogen at intermediate and high energies in the unitarised eikonal-Born series method.' *Journal of Physics B: Atomic and Molecular Physics* 18 (8): 1637.

Byron Jr, F. W., and Charles J. Joachain. 1977. 'Elastic scattering of electrons and positrons by atomic hydrogen and helium at intermediate and high energies.' *Journal of Physics B: Atomic and Molecular Physics* 10 (2): 207.

Böhler, Esther, Jonas Warneke, and Petra Swiderek. 2013. 'Control of chemical reactions and synthesis by low-energy electrons.' *Chemical Society Reviews* 42 (24): 9219-9231.

Cadez, I. M., V. M. Pejcev, and M. V. Kurepa. 1983. 'Electron-sulphur dioxide total ionisation and electron attachment cross sections.' *Journal of Physics D: Applied Physics* 16 (3): 305.

Calogaro, F. 1967. *Variable Phase Approach to Potential Scattering.* New York: Academic Press.

Campbell, L., and M. J. Brunger. 2007. 'Electron impact contribution to infrared NO emissions in auroral conditions.' *Geophysical Research Letters* 34 (22).

———. 2016. 'Electron collisions in atmospheres.' *International Reviews in Physical Chemistry* 35 (2): 297–351.

Campeanu, R. I., R. P. McEachran, and A. D. Stauffer. 2001. 'Positron impact ionization of hydrogen and the noble gases.' *Canadian Journal of Physics* 79 (10): 1231–1236.

Caprasecca, S., J. D. Gorfinkiel, D. Bouchiha, and L. G. Caron. 2009. 'Multiple scattering approach to elastic electron collisions with molecular clusters.' *Journal of Physics B: Atomic, Molecular and Optical Physics* 42 (9): 095205.

Carlson, R. W., M. S. Anderson, R. E. Johnson, W. D. Smythe, A. R. Hendrix, C. A. Barth, L. A. Soderblom et al. 1999. 'Hydrogen peroxide on the surface of Europa.' *Science* 283 (5410): 2062–2064.

Caron, Laurent, D. Bouchiha, J. D. Gorfinkiel, and L. Sanche. 2007. 'Adapting gas-phase electron scattering R-matrix calculations to a condensed-matter environment.' *Physical Review A* 76 (3): 032716.

Carrier, William, Weijun Zheng, Yoshihiro Osamura, and Ralf I. Kaiser. 2006. 'First infrared spectroscopic characterization of digermyl (Ge_2H_5) and d5-digermyl (Ge_2D_5) radicals in low temperature germane matrices.' *Chemical Physics* 325 (2): 499–508.

Cársky, Petr, and Roman Curik (eds). 2011. *Low-Energy Electron Scattering from Molecules, Biomolecules and Surfaces.* Boca Raton: CRC Press.

Celiberto, R., V. Laporta, A. Laricchiuta, J. Tennyson, and J. M. Wadehra. 2014. 'Molecular Physics of Elementary Processes Relevant to Hypersonics: Electron-Molecule Collisions'. *The Open Plasma Physics Journal* 7: 33–47.

Center, R. E., and A. Mandl. 1972. 'Ionization cross sections of F_2 and Cl_2 by electron impact.' *The Journal of Chemical Physics* 57 (10): 4104–4106.

Champion, Christophe. 2013. 'Quantum-mechanical predictions of electron-induced ionization cross sections of DNA components.' *The Journal of Chemical Physics* 138 (18): 184306.

Chatham, H., D. Hils, R. Robertson, and A. Gallagher. 1984. 'Total and partial electron collisional ionization cross sections for CH_4, C_2H_6, SiH_4, and Si_2H_6.' *The Journal of Chemical Physics* 81 (4): 1770–1777.

Chaudhari, Asha S., Foram M. Joshi, and K. N. Joshipura. 2015. 'Eletron impact ionization of plasma important atomic Be, B and BX (X = N, O) targets'. *Prajna - Journal of Pure and Applied Sciences* 22-23: 84–88.

Chiari, Luca, Antonio Zecca, Emanuele Trainotti, Gustavo García, Francisco Blanco, Márcio HF Bettega, Sergio d'A. Sanchez, Márcio T. do N. Varella, Marco AP Lima, and M. J. Brunger. 2013. 'Positron and electron collisions with nitrous oxide: Measured and calculated cross sections.' *Physical Review A* 88 (2): 022708.

Chow, T. P. and A. J. Steckl. 1982. Plasma etching of sputtered Mo and $MoSi_2$ thin films in NF_3 gas mixtures. *Journal of Applied Physics*, 53 (8): 5531–5540.

Christophorou, L. G. (ed.). 2013. *Electron—Molecule Interactions and Their Applications* 2. Academic Press.

Christophorou, Loucas G., James K. Olthoff, and M. V. V. S. Rao. 1996. 'Electron interactions with CF_4.' *Journal of Physical and Chemical Reference Data* 25 (5): 1341–1388/

Christophorou, Loucas G., and James K. Olthoff. 1998. 'Electron Interactions with C_2F_6.' *Journal of Physical and Chemical Reference Data* 27 (1): 1–29.

———. 1999. 'Electron Interactions With Plasma Processing Gases: An Update for CF_4, CHF_3, C_2F_6, and C_3F_8.' *Journal of Physical and Chemical Reference Data* 28 (4): 967–982.

———. 2000. 'Electron interactions with CF_3I.' *Journal of Physical and Chemical Reference Data* 29 (4): 553–569.

———. 2000. 'Electron interactions with SF_6.' *Journal of Physical and Chemical Reference Data* 29 (3): 267–330.

———. 2001. 'Electron collision data for plasma-processing gases.' *Advances in Atomic, Molecular, and Optical Physics* 44: 59–98.

———. 2001. 'Electron interactions with c-C_4F_8.' *Journal of Physical and Chemical Reference Data* 30(2): 449–473.

———. 2012. *Fundamental Electron Interactions with Plasma Processing Gases*. Berlin: Springer Science & Business Media.

Chung, H. K., B. J. Braams, K. Bartschat, A. G. Csaszar, G. W. F. Drake, T. Kirchner, V. Kokoouline, and J. Tennyson. 2016. 'Uncertainty estimates for theoretical atomic and molecular data.' *Journal of Physics D: Applied Physics* 49 (36): 363002.

Clancy, R. T., B. J. Sandor, and G. H. Moriarty-Schieven. 2004. 'A measurement of the 362 GHz absorption line of Mars atmospheric H_2O_2.' *Icarus* 168 (1): 116–121.

Clementi, Enrico, and Carla Roetti. 1974. 'Roothaan-Hartree-Fock atomic wave functions: Basis functions and their coefficients for ground and certain excited states of neutral and ionized atoms, Z≤54'. *Atomic Data and Nuclear Data Tables* 14 (3-4): 177–478.

Coates, A. J., F. J. Crary, G. R. Lewis, D. T. Young, J. H. Waite, and E. C. Sittler. 2007. 'Discovery of heavy negative ions in Titan's ionosphere.' *Geophysical Research Letters* 34 (22).

Colgan, James, S. D. Loch, M. S. Pindzola, C. P. Ballance, and D. C. Griffin. 2003. 'Electron-impact ionization of all ionization stages of beryllium.' *Physical Review A* 68 (3): 032712.

Cosby, P. C. 1993. 'Electron-impact dissociation of nitrogen.' *The Journal of Chemical Physics* 98 (12): 9544–9553.

Cox Jr, H. L., and R. A. Bonham. 1967. 'Elastic electron scattering amplitudes for neutral atoms calculated using the partial wave method at 10, 40, 70, and 100 kV for Z= 1 to Z= 54.' *The Journal of Chemical Physics* 47 (8): 2599–2608.

Čurík, R., P. Čársky, and Michael Allan. 2015. 'Electron-impact vibrational excitation of cyclopropane.' *The Journal of chemical physics* 142 (14): 144312.

da Costa, Romarly F., Fernando J. da Paixão, and Marco AP Lima. 2005. 'Cross sections for electron-impact excitation of the H_2 molecule using the MOB-SCI strategy.' *Journal of Physics B: Atomic, Molecular and Optical Physics* 38 (24): 4363.

Dababneh, M. S., W. E. Kauppila, J. P. Downing, F. Laperriere, V. Pol, J. H. Smart, and T. S. Stein. 1980. 'Measurements of total scattering cross sections for low-energy positrons and electrons colliding with krypton and xenon.' *Physical Review A* 22 (5): 1872.

Dababneh M. S., Y.-F. Hsieh, W. E. Kauppila, C. K. Kwan, Steven J. Smith, T. S. Stein, and M. N. Uddin.1988. 'Total-cross section measurements for positron and electron scattering by O_2, CH_4, and SF_6.' Physical Review A 38(3): 1207

Dababneh, M. S., Y-F. Hsieh, W. E. Kauppila, V. Pol, and T. S. Stein. 1982. 'Total-scattering cross section measurements for intermediate-energy positrons and electrons colliding with Kr and Xe.' *Physical Review A* 26 (3): 1252.

Dalba, G., P. Fornasini, R. Grisenti, G. Ranieri, and A. Zecca. 1980. 'Absolute total cross section measurements for intermediate energy electron scattering. II. N_2, O_2 and NO.' *Journal of Physics B: Atomic and Molecular Physics* 13 (23): 4695.

Dampc, Marcin, Ewelina Szymańska, Brygida Mielewska, and Mariusz Zubek. 2011. 'Ionization and ionic fragmentation of tetrahydrofuran molecules by electron collisions.' *Journal of Physics B: Atomic, Molecular and Optical Physics* 44 (5): 055206.

Danielson, J. R., D. H. E. Dubin, R. G. Greaves, and C. M. Surko. 2015. 'Plasma and trap-based techniques for science with positrons.' *Reviews of Modern Physics* 87 (1): 247.

De Heer, F. J., M. R. C. McDowell, and R. W. Wagenaar. 1977. 'Numerical study of the dispersion relation for e--H scattering.' *Journal of Physics B: Atomic and Molecular Physics* 10 (10): 1945.

De Pablos, J. L., P. A. Kendall, P. Tegeder, A. Williart, F. Blanco, G. García, and N. J. Mason. 2002. 'Total and elastic electron scattering cross sections from ozone at intermediate and high energies.' *Journal of Physics B: Atomic, Molecular and Optical Physics* 35 (4): 865.

Defrance, Pierre, W. Claeys, Alain Cornet, and G. Poulaert. 1981. 'Electron impact ionisation of metastable atomic hydrogen.' *Journal of Physics B: Atomic and Molecular Physics* 14 (1): 111.

Deutsch, H., K. Becker, R. Basner, M. Schmidt, and T. D. Märk. 1998. 'Application of the modified additivity rule to the calculation of electron-impact ionization cross sections of complex molecules.' *The Journal of Physical Chemistry A* 102 (45): 8819–8826.

Deutsch, H., K. Becker, R. K. Janev, M. Probst, and T. D. Märk. 2000. 'Isomer effect in the total electron impact ionization cross section of cyclopropane and propene (C_3H_6).' *Journal of Physics B: Atomic, Molecular and Optical Physics* 33 (24): L865.

Deutsch, H., K. Becker, S. Matt, and T. D. Märk. 2000. 'Theoretical determination of absolute electron-impact ionization cross sections of molecules.' *International Journal of Mass Spectrometry* 197 (1): 37–69.

Deutsch, H., T. D. Maerk, V. Tarnovsky, K. Becker, C. Cornelissen, L. Cespiva, and V. Bonacic-Koutecky. 1994. 'Measured and calculated absolute total cross sections for the single ionization of CFx and NFx by electron impact'. *International Journal of Mass Spectrometry and Ion Processes* 137 (C): 77–91.

Deutsch, H., K. Becker, and T. D. Märk. 2000. 'Calculation of absolute electron-impact ionization cross sections of dimers and trimers.' *The European Physical Journal D-Atomic, Molecular, Optical and Plasma Physics* 12 (2): 283–287.

Deutsch, H., K. Hilpert, K. Becker, M. Probst, and T. D. Märk. 2001. 'Calculated absolute electron-impact ionization cross sections for AlO, Al_2O, and WO_x (x= 1–3).' *Journal of Applied Physics* 89 (3): 1915–1921.

Dewangan, D. P. 2012. 'Asymptotic methods for Rydberg transitions.' *Physics Reports* 511 (1): 1–142.

Dixon, A. J., A. Von Engel, and M. F. A. Harrison. 1975. 'A measurement of the electron impact ionization cross section of atomic hydrogen in the metastable 2s state.' *Proceedings of the Royal Society of London A: Mathematical, Physical and Engineering Sciences* 343 (1634): 333–349.

Dixon, A. J., M. F. A. Harrison, and A. C. H. Smith. 1976. 'A measurement of the electron impact ionization cross section of helium atoms in metastable states.' *Journal of Physics B: Atomic and Molecular Physics* 9 (15): 2617.

Djuric, N., I. Cadez, and M. Kurepa. 1989. 'Total Electron Impact Ionization Cross Sections for Methanol, Ethanol and n-Propanol Molecules'. *Fizika* 21: 339–343.

Dmitrieva, I. K., and G. I. Plindov. 1983. 'Dipole polarizability, radius and ionization potential for atomic systems.' *Physica Scripta* 27 (6): 402.

Donnelly, Vincent M., Daniel L. Flamm, W. C. Dautremont-Smith, and D. J. Werder. 1984. 'Anisotropic etching of SiO_2 in low-frequency CF_4/O_2 and NF_3/Ar plasmas.' *Journal of Applied Physics* 55 (1): 242–252.

Drexel, H., G. Senn, T. Fiegele, P. Scheier, A. Stamatovic, N. J. Mason, and T. D. Märk. 2001. 'Dissociative electron attachment to hydrogen'. *Journal of Physics B: Atomic Molecular and Optical Physics* 34 (8): 1415–1424.

DuBois, R. D., and M. E. Rudd. 1976. 'Differential cross sections for elastic scattering of electrons from argon, neon, nitrogen and carbon monoxide.' *Journal of Physics B: Atomic and Molecular Physics* 9 (15): 2657.

Ehhalt, D. H., H-P. Dorn, and D. Poppe. 1990. 'The chemistry of the hydroxyl radical in the troposphere.' *Proceedings of the Royal Society of Edinburgh, Section B: Biological Sciences* 97: 17–34.

Encrenaz, Th, Bruno Bézard, T. K. Greathouse, M. J. Richter, J. H. Lacy, S. K. Atreya, A. S. Wong, Sébastien Lebonnois, Franck Lefèvre, and François Forget. 2004. 'Hydrogen peroxide on Mars: evidence for spatial and seasonal variations.' *Icarus* 170 (2): 424–429.

Engmann, Sarah, Michal Stano, Peter Papp, Michael J. Brunger, Štefan Matejčík, and Oddur Ingólfsson. 2013. 'Absolute cross sections for dissociative electron attachment and dissociative ionization of cobalt tricarbonyl nitrosyl in the energy range from 0 eV to 140 eV.' *The Journal of Phemical Physics* 138 (4): 044305.

Fabrikant, Ilya I. 2016. 'Long-range effects in electron scattering by polar molecules.' *Journal of Physics B: Atomic, Molecular and Optical Physics* 49 (22): 222005.

Fabrikant, Ilya I., S. Caprasecca, Gordon A. Gallup, and J. D. Gorfinkiel. 2012. 'Electron attachment to molecules in a cluster environment.' *The Journal of chemical physics* 136 (18): 184301.

Fabrikant, Ilya I., Samuel Eden, Nigel J. Mason, and Juraj Fedor. 2017. 'Recent Progress in Dissociative Electron Attachment: From Diatomics to Biomolecules.' *Advances in Atomic, Molecular, and Optical Physics* 66: 545.

Fayer, S. E., S. J. Brawley, M. Shipman, L. Sarkadi, and G. Laricchia. 2015. 'Positronium beam production and scattering at low energies.' *Journal of Physics: Conference Series* 635 (5): 052069.

Feldman, Paul D., Michael F. A'Hearn, Jean-Loup Bertaux, Lori M. Feaga, Joel Wm Parker, Eric Schindhelm, Andrew J. Steffl et al. 2015. 'Measurements of the near-nucleus coma of comet 67P/Churyumov-Gerasimenko with the Alice far-ultraviolet spectrograph on Rosetta.' *Astronomy & Astrophysics* 583: A8.

Feldman, Paul D., Michael F. A'hearn, Lori M. Feaga, Jean-Loup Bertaux, John Noonan, Joel Wm Parker, Eric Schindhelm, Andrew J. Steffl, S. Alan Stern, and Harold A. Weaver. 2016. 'The nature and frequency of the gas outbursts in comet 67P/Churyumov–Gerasimenko observed by the ALICE far-ultraviolet spectrograph on Rosetta.' *The Astrophysical Journal Letters* 825 (1): L8.

Field, D., D. W. Knight, G. Mrotzek, J. Randell, S. L. Lunt, J. B. Ozenne, and J. P. Ziesel. 1991. 'A high-resolution synchrotron photoionization spectrometer for the study of low-energy electron-molecule scattering.' *Measurement Science and Technology* 2 (8): 757.

Floeder, K., D. Fromme, W. Raith, A. Schwab, and G. Sinapius. 1985. 'Total cross section measurements for positron and electron scattering on hydrocarbons between 5 and 400 eV.' *Journal of Physics B: Atomic and Molecular Physics* 18 (16): 3347.

Francis-Staite, Jessica R., Brett A. Schmerl, Michael J. Brunger, H. Kato, and Stephen J. Buckman. 2010. 'Elastic electron scattering from CF_3I.' *Physical Review A* 81 (2): 022704.

Franck, James, and Gustav Hertz. 1914. 'Über Zusammenstöße zwischen Elektronen und Molekülen des Quecksilberdampfes und die Ionisierungsspannung desselben' (in German) [On the collisions between electrons and molecules of mercury vapour and the ionization potential of the same] *Verhandlungen der Deutschen Physikalischen Gesellschaft* 16: 457–467.

Franz, Gerhard. 2009. *Low Pressure Plasmas and Micro-structuring Technology*. New York: Springer Science & Business Media.

Freund, Robert S., Robert C. Wetzel, Randy J. Shul, and Todd R. Hayes. 1990. 'Cross section measurements for electron-impact ionization of atoms.' *Physical Review A* 41 (7): 3575.

Fricke, B. 1986. 'On the correlation between electric polarizabilities and the ionization potential of atoms.' *The Journal of Chemical Physics* 84 (2): 862–866.

Fridman, Alexander, and Gary Friedman. 2013. *Plasma Medicine*. Hoboken, New Jersey: John Wiley and Sons Inc.

Fujimoto, M. M., and M. T. Lee. 2000. 'Elastic and absorption cross sections for electron-nitric oxide collisions.' *Journal of Physics B: Atomic, Molecular and Optical Physics* 33 (21): 4759.

Fursa, Dmitry V., and Igor Bray. 1997. 'Convergent close-coupling calculations of electron-beryllium scattering.' *Journal of Physics B: Atomic, Molecular and Optical Physics* 30 (8): L273.

———. 1997. 'Convergent close-coupling calculations of electron scattering on helium-like atoms and ions: electron-beryllium scattering.' *Journal of Physics B: Atomic, Molecular and Optical Physics* 30 (24): 5895.

Fuss, M., A. Muñoz, J. C. Oller, F. Blanco, D. Almeida, P. Limão-Vieira, T. P. D. Do, M. J. Brunger, and G. García. 2009. 'Electron-scattering cross sections for collisions with tetrahydrofuran from 50 to 5000 eV.' *Physical Review A* 80 (5): 052709.

Galand, Marina, Luke Moore, Ingo Mueller-Wodarg, Michael Mendillo, and Steve Miller. 2011. 'Response of Saturn's auroral ionosphere to electron precipitation: Electron density, electron temperature, and electrical conductivity.' *Journal of Geophysical Research: Space Physics* 116 (A9).

Gangopadhyay, Sumona. 2008. 'Electron impact atomic molecular processes relevant in planetary and astrophysical systems – a theoretical study', Ph. D. thesis (unpublished), Sardar Patel University, Vallabh Vidyanagar, India. p.203.

Ganguly, P. 2008. 'Atomic sizes and atomic properties.' *Journal of Physics B: Atomic, Molecular and Optical Physics* 41 (10): 105002.

Garcia, G., C. Aragon, and J. Campos. 1990.' Total cross sections for electron scattering from CO in the energy range 380–5200 eV.' *Physical Review A* 42 (7): 4400.

García, Gómez-Tejedor, Gustavo, Fuss, and Martina Christina Fuss (eds). 2012. 'Radiation Damage in Biomolecular Systems.' Accessed 1 March 2018. doi: https://doi.org/10.1007/978-94-007-2564-5

Garcia, G., A. Perez, and J. Campos. 1988. 'Total cross section for electron scattering from N_2 in the energy range 600–5000 eV'. *Physical Review A* 38 (2): 654.

Gaudin, A., and R. J. Hagemann. 1967. 'Absolute determination of the total and partial effective ionization cross sections of helium, neon, argon, and acetylene for 100–2000 eV electrons.' *Journal de Chimie Physique* 64: 1209–1221.

Gay, T. J. 2009. 'Physics and technology of polarized electron scattering from atoms and molecules.' *Advances in Atomic, Molecular, and Optical Physics* 57: 157–247.

Gedeon, Viktor, Sergej Gedeon, Vladimir Lazur, Elizabeth Nagy, Oleg Zatsarinny, and Klaus Bartschat. 2014. 'B-spline R-matrix-with-pseudostates calculations for electron-impact excitation and ionization of fluorine.' *Physical Review A* 89 (5): 052713.

Geller, Richard. 1996. *Electron Cyclotron Resonance Ion Sources and ECR Plasmas.* Boca Raton: CRC Press.

Geppert, Wolf Dietrich, and Mats Larsson. 2008. 'Dissociative recombination in the interstellar medium and planetary ionospheres.' *Molecular Physics* 106 (16-18): 2199–2226.

Gergely, G., A. Konkol, M. Menyhard, B. Lesiak, A. Jablonski, D. Varga, and J. Toth. 1997. 'Determination of the inelastic mean free path (IMFP) of electrons in germanium and silicon by elastic peak electron spectroscopy (EPES) using an analyser of high resolution.' *Vacuum* 48 (7-9): 621–624.

Gianturco, F. A., and P. Gori Giorgi. 1996. 'Radiative association of LiH $(X\,^1\Sigma^+)$ from electronically excited lithium atoms.' *Physical Review A* 54 (5): 4073.

Goswami, Biplab, Ujjval Saikia, Rahla Naghma, and Bobby Antony. 2013. 'Electron impact ionization cross sections for plasma wall coating elements'. *Chinese Journal of Physics* 51 (6): 1172–1183.

Goswami, Biplab, Rahla Naghma, and Bobby Antony. 2013. 'Cross sections for electron collisions with NF_3.' *Physical Review A* 88 (3): 032707.

Goswami, Biplab, and Bobby Antony. 2014. 'Electron impact scattering by SF_6 molecule over an extensive energy range.' *RSC Advances* 4 (58): 30953–30962.

———. 2014. 'On the electron impact ionization of silicon and metal containing organic molecules.' *International Journal of Mass Spectrometry* 361: 28–33.

Gradshteyn, I. S., and I. M. Ryzhik. 1965. 'Table of integrals, series, and products prepared by Ju. V. Geronimus and M. Ju. Ceıtlin.' Translated from the Russian by Scripta Technica, Inc. Translation edited by Alan Jeffrey.

Gribakin, G. F., Jason Asher Young, and C. M. Surko. 2010. 'Positron-molecule interactions: Resonant attachment, annihilation, and bound states.' *Reviews of Modern Physics* 82 (3): 2557.

Gries, Werner H. 1996. 'A Universal Predictive Equation for the Inelastic Mean Free Pathlengths of X-ray Photoelectrons and Auger Electrons.' *Surface and Interface Analysis* 24 (1): 38–50.

Griffith, T. C., G. R. Heyland, K. S. Lines, and T. R. Twomey. 1979. 'Total cross sections for the scattering of positrons by helium, neon, and argon at intermediate energies.' *Applied Physics A: Materials Science & Processing* 19 (4): 431–437.

Grodent, Denis, J. Hunter Waite, and Jean-Claude Gérard. 2001. 'A self-consistent model of the Jovian auroral thermal structure.' *Journal of Geophysical Research: Space Physics* 106 (A7): 12933–12952.

Guberman, Steven L. (ed.). 2012. *Dissociative recombination of molecular ions with electrons.* Heidelberg: Springer Science & Business Media.

Gupta, Dhanoj, and Bobby Antony. 2014. 'Electron impact ionization of cycloalkanes, aldehydes, and ketones.' *The Journal of Chemical Physics* 141 (5): 054303.

Gupta, Dhanoj, Heechol Choi, Mi-Young Song, Grzegorz P. Karwasz, and Jung-Sik Yoon. 2017. 'Electron impact ionization cross section studies of C_2F_x (x= 1– 6) and C_3F_x (x= 1– 8) fluorocarbon species.' *The European Physical Journal D* 71 (4): 88.

Gupta, Dhanoj, Rahla Naghma, and Bobby Antony. 2013. 'Electron impact total and ionization cross sections for Sr, Y, Ru, Pd, and Ag atoms.' *Canadian Journal of Physics* 91 (9): 744– 750.

Gurung, Meera Devi, and W. M. Ariyasinghe. 2017. 'Total electron scattering cross sections of some important biomolecules at 0.2–6.0 keV energies.' *Radiation Physics and Chemistry.* 141: 01

Gutkin, M., J. M. Mahoney, V. Tarnovsky, H. Deutsch, and Kurt Becker. 2009. 'Electron-impact ionization of the $SiCl_3$ radical.' *International Journal of Mass Spectrometry* 280 (1): 101– 106.

Gérard, J-C., Bertrand Bonfond, Denis Grodent, Aikaterini Radioti, J. T. Clarke, G. R. Gladstone, J. H. Waite, D. Bisikalo, and V. I. Shematovich. 2014. 'Mapping the electron energy in Jupiter's aurora: Hubble spectral observations.' *Journal of Geophysical Research: Space Physics* 119 (11): 9072–9088.

Gómez-Tejedor, Gustavo García, and Martina Christina Fuss (eds). 2012. *Radiation Damage in Biomolecular Systems.* Heidelberg: Springer Science & Business Media.

Haaland, P. D., C. Q. Jiao, and A. Garscadden. 2001. 'Ionization of NF_3 by electron impact'. *Chemical Physics Letters* 340 (5-6): 479–483.

Haider, S. A., S. M. P. McKenna-Lawlor, C. D. Fry, Rajmal Jain, and K. N. Joshipura. 2012. 'Effects of solar X-ray flares in the E region ionosphere of Mars: First model results.' *Journal of Geophysical Research: Space Physics* 117 (A5).

Haider, S. A., Varun Sheel, and Shyam Lal (eds). 2010. *Modeling of Planetary Atmospheres.* Delhi: Macmillan Publishers India Ltd.

Hamilton, James R., Jonathan Tennyson, Shuo Huang, and Mark J. Kushner. 2017. 'Calculated cross sections for electron collisions with NF_3, NF_2 and NF with applications to remote plasma sources.' *Plasma Sources Science and Technology* 26 (6): 065010.

Hara, Shunsuke. 1967. 'The scattering of slow electrons by hydrogen molecules.' *Journal of the Physical Society of Japan* 22 (3): 710–718.

Hargreaves, L. R., J. R. Brunton, A. Prajapati, Masamitsu Hoshino, Francisco Blanco, Gustavo Garcia, S. J. Buckman, and Michael J. Brunger. 2011. 'Elastic cross sections for electron scattering from iodomethane.' *Journal of Physics B: Atomic, Molecular and Optical Physics* 44 (4): 045207.

Hargreaves, L., K. Ralphs, G. Serna, M. A. Khakoo, C. Winstead, and V. McKoy. 2012. 'Excitation of the a^3B_1 and A^1B_1 states of H_2O by low-energy electron impact.' *Journal of Physics B: Atomic, Molecular and Optical Physics* 45 (20): 201001.

Harland, Peter W., and Claire Vallance. 1997. 'Ionization cross sections and ionization efficiency curves from polarizability volumes and ionization potentials'. *International Journal of Mass Spectrometry and Ion Processes* 171 (1-3): 173–181.

Harnisch, Jochen, Matthias Frische, Reinhard Borchers, Anton Eisenhauer, and Armin Jordan. 2000. 'Natural fluorinated organics in fluorite and rocks.' *Geophysical Research Letters* 27 (13): 1883–1886.

Haume, Kaspar, Soraia Rosa, Sophie Grellet, Małgorzata A. Śmiałek, Karl T. Butterworth, Andrey V. Solov'yov, Kevin M. Prise, Jon Golding, and Nigel J. Mason. 2016. 'Gold nanoparticles for cancer radiotherapy: a review.' *Cancer Nanotechnology* 7 (1): 8.

Hayashi, Makoto. 1987. 'Boltzmann Equation Analysis of the Synergism of Uniform Field Breakdown Voltage for SF_6-CCl_2F_2 and SF_6-SO_2 Mixtures'. *Swarm Studies and Inelastic Electron-Molecule Collisions*. New York: Plenum Press. pp 101.

Hayashi, M., M. Capitelli, and J. N. Bardsley (eds). 1990. *Non-equilibrium Processes in Partially Ionized Gases*. New York: Plenum Press. pp 333.

Hayes, Todd R., Robert C. Wetzel, and Robert S. Freund. 1987. 'Absolute electron-impact-ionization cross section measurements of the halogen atoms.' *Physical Review A* 35 (2): 578.

Hermann, Andreas, and Peter Schwerdtfeger. 2011. 'Blueshifting the onset of optical UV absorption for water under pressure.' *Physical Review Letters* 106 (18): 187403.

Herrmann, D., K. Jost, J. Kessler, and M. Fink. 1976. 'Differential cross sections for elastic electron scattering. II. Charge cloud polarization in N_2.' *The Journal of Chemical Physics* 64 (1): 1–5.

Hirota, Tomoya, Satoshi Yamamoto, Hitomi Mikami, and Masatoshi Ohishi. 1998. 'Abundances of HCN and HNC in dark cloud cores.' *The Astrophysical Journal* 503 (2): 717.

Hoffman, K. R., M. S. Dababneh, Y-F. Hsieh, W. E. Kauppila, V. Pol, J. H. Smart, and T. S. Stein. 1982. 'Total-cross section measurements for positrons and electrons colliding with H_2, N_2, and CO_2.' *Physical Review A* 25 (3): 1393.

Hudson, J. E., M. L. Hamilton, C. Vallance, and P. W. Harland. 2003. 'Absolute electron impact ionization cross sections for the C_1 to C_4 alcohols.' *Physical Chemistry Chemical Physics* 5 (15): 3162–3168.

Hudson, James E., Claire Vallance, Mark Bart, and Peter W. Harland. 2001. 'Absolute electron-impact ionization cross sections for a range of C_1 to C_5 chlorocarbons.' *Journal of Physics B: Atomic, Molecular and Optical Physics* 34 (15): 3025.

Huo, Winifred M., Vladimir Tarnovsky, and Kurt H. Becker. 2002. 'Total electron-impact ionization cross sections of CF_x and NF_x (x= 1–3).' *Chemical Physics Letters* 358 (3): 328–336.

Hwang, W., Y-K. Kim, and M. Eugene Rudd. 1996. 'New model for electron-impact ionization cross sections of molecules.' *The Journal of Chemical Physics* 104 (8): 2956–2966.

Iga, I., M. V. V. S. Rao, and S. K. Srivastava. 1996. 'Absolute electron impact ionization cross sections for N_2O and NO from threshold up to 1000 eV.' *Journal of Geophysical Research: Planets* 101 (E4): 9261–9266.

Itikawa Yukikazu., 1971. 'Effective collision frequency of electrons in atmospheric gases', *Planetary and Space Science* 19 (8); 993–1007.

———. 1978. 'Electron scattering by polar molecules.' *Physics Reports* 46 (4): 117–164.

———. 2002. 'Cross sections for electron collisions with carbon dioxide.' *Journal of Physical and Chemical Reference Data* 31 (3): 749–767.

———. 2006. 'Cross sections for electron collisions with nitrogen molecules.' *Journal of Physical and Chemical Reference Data* 35 (1): 31–53.

———. 2009. 'Cross sections for electron collisions with oxygen molecules.' *Journal of Physical and Chemical Reference Data* 38 (1): 1–20.

———. 2015. 'Cross Sections for Electron Collisions with Carbon Monoxide.' *Journal of Physical and Chemical Reference Data* 44 (1): 013105.

———. 2016. 'Cross sections for electron collisions with nitric oxide.' *Journal of Physical and Chemical Reference Data* 45 (3): 033106.

Itikawa, Yukikazu (ed.). 2003. *Photon and Electron Interactions with Atoms, Molecules and Ions.* New York: Springer.

Itikawa, Y., M. Hayashi, A. Ichimura, K. Onda, K. Sakimoto, K. Takayanagi, M. Nakamura, H. Nishimura, and T. Takayanagi. 1986. 'Cross sections for collisions of electrons and photons with nitrogen molecules'. *Journal of Physical and Chemical Reference Data* 15 (3): 985–1010.

Itikawa, Y., and A. Ichimura. 1990. 'Cross sections for collisions of electrons and photons with atomic oxygen.' *Journal of Physical and Chemical Reference Data* 19 (3): 637–651.

Itikawa, Yukikazu, and Nigel Mason. 2005. 'Cross sections for electron collisions with water molecules.' *Journal of Physical and Chemical reference data* 34 (1): 1–22.

Jacobsen, F. M., N. P. Frandsen, H. Knudsen, U. Mikkelsen, and D. M. Schrader. 1995. 'Single ionization of He, Ne and Ar by positron impact.' *Journal of Physics B: Atomic, Molecular and Optical Physics* 28 (21): 4691.

Jain, Ashok, and K. L. Baluja. 1992. 'Total (elastic plus inelastic) cross sections for electron scattering from diatomic and polyatomic molecules at 10–5000 eV: H_2, Li_2, HF, CH_4, N_2, CO, C_2H_2, HCN, O_2, HCl, H_2S, PH_3, SiH_4, and CO_2.' *Physical Review A* 45 (1):202.

Jain, Ashok, B. Etemadi, and K. R. Karim. 1990. 'Total (elastic plus inelastic) cross sections for electron scattering with argon and krypton atoms at energies 10-6000 eV.' *Physica Scripta* 41 (3): 321.

Jain, A., A. N. Tripathi, and M. K. Srivastava. 1979. 'Elastic scattering of electrons by molecular hydrogen at intermediate and high energies.' *Physical Review A* 20 (6): 2352–2355.

Janev, Ratko K., William D. Langer, and E. Douglass Jr. 2012. '*Elementary processes in hydrogen-helium plasmas: cross sections and reaction rate coefficients.*' Berlin: Springer.

Jhanwar, B. L., S. P. Khare, and M. K. Sharma. 1980. 'Exchange effects in the independent-atom model for electron-hydrogen-molecule elastic scattering.' *Physical Review A* 22 (6): 2451.

Jheeta, Sohan, A. Domaracka, S. Ptasinska, B. Sivaraman, and N. J. Mason. 2013. 'The irradiation of pure CH_3OH and 1: 1 mixture of NH_3: CH_3OH ices at 30K using low energy electrons.' *Chemical Physics Letters* 556: 359–364.

Jheeta, S., S. Ptasinska, B. Sivaraman, and N. J. Mason. 2012. 'The irradiation of 1: 1 mixture of ammonia: carbon dioxide ice at 30K using 1kev electrons.' *Chemical Physics Letters* 543: 208–212.

Jiang, Yuhai, Jinfeng Sun, and Linde Wan. 1995. 'Total cross sections for electron scattering by polyatomic molecules at 10–1000 eV: H_2S, SiH_4, CH_4, CF_4, CCl_4, SF_6, C_2H_4, CCl_3F, $CClF_3$, and CCl_2F_2.' *Physical Review A* 52 (1): 398.

Jiang, Yuhai, Jinfeng Sun, and Lingde Wan. 2000. 'Additivity rule for the calculation of electron scattering from polyatomic molecules.' *Physical Review A* 62 (6): 062712.

Jiao, C. Q., S. F. Adams, and A. Garscadden. 2009. 'Ionization of 2, 5-dimethylfuran by electron impact and resulting ion-parent molecule reactions.' *Journal of Applied Physics* 106 (1): 013306.

Jiao, C. Q., B. Ganguly, C. A. DeJoseph, and A. Garscadden. 2001. 'Comparisons of electron impact ionization and ion chemistries of CF_3Br and CF_3I.' *International Journal of Mass Spectrometry* 208 (1): 127–133.

Jiao, C. Q., A. Garscadden, and P. D. Haaland. 1998. 'Ion chemistry in octafluorocyclobutane, c-C_4F_8.' *Chemical Physics Letters* 297 (1): 121–126.

———. 1999. 'Partial ionization cross sections of C_2F_6.' *Chemical Physics Letters* 310 (1): 52–56.

Joachain, Charles Jean. 1983. *Quantum Collision Theory*. New York. North-Holland.

Johnson III, Russell D. 1999. 'NIST 101. Computational Chemistry Comparison and Benchmark Database.' *CCCBDB Computational Chemistry Comparison and Benchmark Database.*

Johnson, P. V., I. Kanik, D. E. Shemansky, and X. Liu. 2003. 'Electron-impact cross sections of atomic oxygen.' *Journal of Physics B: Atomic, Molecular and Optical Physics* 36 (15): 3203.

Jones, D. B., R. F. da Costa, MT do N. Varella, M. H. F. Bettega, M. A. P. Lima, F. Blanco, G. García, and M. J. Brunger. 2016. 'Integral elastic, electronic-state, ionization, and total cross sections for electron scattering with furfural.' *The Journal of Chemical Physics* 144 (14): 144303.

Joshi, Foram M. 2013. *Electron Impact Cross sections of Targets Relevant to Plasma and Astrophysics*. Germany: LAP Lambert Academic Publishing.

Joshi, Foram M., and K. N. Joshipura. 2015. 'Scattering of electrons with atomic Mo: free and metallic phases.' *Journal of Physics: Conference Series*, 635 (5): 052081.

Joshi, Foram M., K. N. Joshipura, Asha S. Chaudhari, Hitesh S. Modi, and Manish J. Pindaria. 2016. 'Completing electron scattering studies with the inert gas column: e-Rn scattering and Ionization.' *arXiv preprint arXiv:1602.01928.*

———. 2016. 'Electron impact ionization in plasma technologies; studies on atomic boron and BN molecule.' *AIP Conference Proceedings* 1728 (1): 020186.

Joshipura, K. N. 1989. 'Elastic scattering of electrons from oxygen and ozone molecules at high energies.' *Pramana* 32 (2): 139–142.

———. 2013. *Resonance – Journal of Science Education* 18 (9): 799.

Joshipura, K. N., and B. K. Antony. 2001. 'Total (including ionization) cross sections of electron impact on ground state and metastable Ne atoms.' *Physics Letters A* 289 (6): 323–328.

Joshipura, K. N., B. K. Antony, and Minaxi Vinodkumar. 2002. 'Electron scattering and ionization of ozone, O_2 and O_4 molecules.' *Journal of Physics B: Atomic, Molecular and Optical Physics* 35 (20): 4211.

Joshipura, K. N., and H. S. Desai. 1980. 'Application of Glauber approximation to Feshbach-type resonance scattering.' *Indian Journal of Pure and Applied Physics* 18 (8): 615–617.

Joshipura, K. N., and Sumona Gangopadhyay. 2008. 'Electron collisions with sulfur compounds SO, SO_2 and SO_2AB (A, B = Cl, F): various total cross sections.' *Journal of Physics B: Atomic, Molecular and Optical Physics* 41 (21): 215205.

Joshipura, K. N., Sumona Gangopadhyay, Harshit N. Kothari, and Foram A. Shelat. 2009. 'Calculations on various total cross sections of electron impact on group VA–atoms-threshold to 2000 eV.' In *Journal of Physics: Conference Series* 194 (4): 042038.

———. 2009. 'Total electron scattering and ionization of N, N_2 and metastable excited N_2^*: Theoretical cross sections.' *Physics Letters A* 373 (32): 2876–2881.

Joshipura, K. N., Sumona Gangopadhyay, C. G. Limbachiya, and Minaxi Vinodkumar. 2007. 'Electron impact ionization of water molecules in ice and liquid phases.' *Journal of Physics: Conference Series* 80 (1): 012008.

Joshipura, K. N., Sumona Gangopadhyay, and Bhushit G. Vaishnav. 2006. 'Electron scattering and ionization of NO, N_2O, NO_2, NO_3 and N_2O_5 molecules: theoretical cross sections.' *Journal of Physics B: Atomic, Molecular and Optical Physics* 40 (1): 199.

Joshipura, K. N., Harshit N. Kothari, Foram A. Shelat, Pooja Bhowmik, and N. J. Mason. 2010. 'Electron scattering with metastable H_2^* ($c^3\pi_u$) molecules: ionization and other total cross sections.' *Journal of Physics B: Atomic, Molecular and Optical Physics* 43 (13): 135207.

Joshipura, K. N., and Chetan G. Limbachiya. 2002. 'Theoretical total ionization cross sections for electron impact on atomic and molecular halogens.' *International Journal of Mass Spectrometry* 216 (3): 239–247.

Joshipura, Kamalnayan N., Siddharth H. Pandya, and Nigel J. Mason. 2017. 'Electron scattering and ionization of H_2O: OH, H_2O_2, HO_2 radicals and $(H_2O)_2$ dimer.' *The European Physical Journal D* 71 (4): 96.

Joshipura, K. N., and P. M. Patel. 1993. 'Cross sections of e–-O scattering at intermediate and high energies ($E_i = 8.7$–1000 eV).' *Physical Review A* 48 (3): 2464.

———. 1994. 'Electron impact total (elastic+ inelastic) cross sections of C, N & O atoms and their simple molecules.' *Zeitschrift für Physik D Atoms, Molecules and Clusters* 29 (4): 269–273.

Joshipura, K. N., B. G. Vaishnav, and Sumona Gangopadhyay. 2007. 'Electron impact ionization cross sections of plasma relevant and astrophysical silicon compounds: SiH_4, Si_2H_6, $Si(CH_3)_4$, SiO, SiO_2, SiN and SiS.' *International Journal of Mass Spectrometry* 261 (2): 146–151.

Joshipura, K. N., B. G. Vaishnav, and C. G. Limbachiya. 2006. 'Ionization and excitation of some atomic targets and metal oxides by electron impact.' *Pramana* 66 (2): 403–414.

Joshipura, K. N., and Minaxi Vinodkumar. 1997. 'Total cross sections of electron collisions with S atoms; H_2S, OCS and SO_2 molecules ($E_i \geq 50$ eV).' *Zeitschrift für Physik D Atoms, Molecules and Clusters* 41 (2): 133–137.

———. 1999. 'Various total cross sections for electron impact on C_2H_2, C_2H_4 and CH_3X (X=CH_3,OH,F,NH_2).' *The European Physical Journal D-Atomic, Molecular, Optical and Plasma Physics* 5 (2): 229–235.

Joshipura, K. N., Minaxi Vinodkumar, C. G. Limbachiya, and B. K. Antony. 2004. 'Calculated total cross sections of electron-impact ionization and excitations in tetrahedral (XY_4) and SF_6 molecules.' *Physical Review A* 69 (2): 022705.

Joshipura, K. N., Minaxi Vinodkumar, and Umesh M. Patel. 2001. 'Electron impact total cross sections of CHx, NHx and OH radicals vis-à-vis their parent molecules.' *Journal of Physics B: Atomic, Molecular and Optical Physics* 34 (4): 509.

Kanik, I., J. C. Nickel, and S. Trajmar. 1992. 'Total electron scattering cross section measurements for Kr, O_2 and CO.' *Journal of Physics B: Atomic, Molecular and Optical Physics* 25 (9): 2189.

Kanik, I., S. Trajmar, and J. C. Nickel. 1992. 'Total cross section measurements for electron scattering on CH_4 from 4 to 300 eV.' *Chemical Physics Letters* 193 (4): 281–286.

Kara, V., K. Paludan, J. Moxom, P. Ashley, and G. Laricchia. 1997. 'Single and double ionization of neon, krypton and xenon by positron impact.' *Journal of Physics B: Atomic, Molecular and Optical Physics* 30 (17): 3933.

Karim, K. R., and Ashok Jain. 1989. 'Elastic scattering of electrons from argon atoms at 0.001-300 eV.' *Physica Scripta* 39 (2): 238.

Karwasz, G., P. Mozejko, and Mi-Young Song. 2014. 'Electron-impact ionization of fluoromethanes – Review of experiments and binary-encounter models'. *International Journal of Mass Spectrometry* 365/366: 232–237.

Karwasz, G., R. S. Brusa, A. Gasparoli, and A. Zecca. 1993. 'Total cross section measurements for e⁻–CO scattering: 80–4000 eV.' *Chemical physics letters* 211 (6): 529–533.

Karwasz, Grzegorz P., Mario Barozzi, Roberto S. Brusa, and Antonio Zecca. 2002. 'Total cross sections for positron scattering on argon and krypton at intermediate and high energies.' *Nuclear Instruments and Methods in Physics Research Section B: Beam Interactions with Materials and Atoms* 192 (1): 157–161.

Karwasz, Grzegorz P., Roberto S. Brusa, and Antonio Zecca. 2001. 'One century of experiments on electron-atom and molecule scattering: a critical review of integral cross sections.' *Riv Nuovo Cimento* 24 (1): 1–146.

Kauppila, W. E., T. S. Stein, J. H. Smart, M. S. Dababneh, Y. K. Ho, J. P. Downing, and V. Pol. 1981. 'Measurements of total scattering cross sections for intermediate-energy positrons and electrons colliding with helium, neon, and argon.' *Physical Review A* 24 (2): 725.

Kaur, Gurpreet, Arvind Kumar Jain, Harsh Mohan, Parjit S. Singh, Sunita Sharma, and A. N. Tripathi. 2015. 'Studies of cross sections for collisions of electrons from hydride molecules: NH_3 and PH_3.' *Physical Review A* 91 (2): 022702.

Kaur, Jaspreet, and Bobby Antony. 2015. 'Electron induced ionization cross sections for astrophysical modelling'. *International Journal of Mass Spectrometry* 386: 24–31.

Kaur, Jaspreet, Dhanoj Gupta, Rahla Naghma, Debdeep Ghoshal, and Bobby Antony. 2014. 'Electron impact ionization cross sections of atoms.' *Canadian Journal of Physics* 93 (6): 617–625.

Khakoo, Murtadha A., John Muse, Kevin Ralphs, Romarly F. da Costa, Márcio HF Bettega, and Marco AP Lima. 2010. 'Low-energy elastic electron scattering from furan.' *Physical Review A* 81 (6): 062716.

Khakoo, M. A., and Sandor Trajmar. 1986. 'Elastic electron scattering cross sections for molecular hydrogen.' *Physical Review A* 34 (1): 138.

Khakoo, M. A., H. Silva, J. Muse, M. C. A. Lopes, C. Winstead, and V. McKoy. 2008. 'Electron scattering from H_2O: elastic scattering.' *Physical Review A* 78 (5): 052710.

Khakoo, M. A., P. Vandeventer, J. G. Childers, I. Kanik, C. J. Fontes, K. Bartschat, V. Zeman et al. 2004. 'Electron impact excitation of the argon $3p^54s$ configuration: differential cross sections and cross section ratios.' *Journal of Physics B: Atomic, Molecular and Optical Physics* 37 (1): 247.

Khare S. P. 2001. *Introduction to the Theory of Collisions of Electrons with Atoms and Molecules.* New York: Kluwer Academic/Plenum Publishers. *See also*, Khare, Satya Prakash. 2012. '*Introduction to the Theory of Collisions of Electrons with Atoms and Molecules.*' New York: Springer Science & Business Media.

Khare, S. P., and K. Lata. 1985. 'Elastic scattering of electrons and positrons by the helium atom and the hydrogen molecule at intermediate energies'. *Journal of Physics B: Atomic Molecular and Optical Physics* 18 (14): 2941–2954.

Khare, S. P., M. K. Sharma, and Surekha Tomar. 1999. 'Electron impact ionization of methane.' *Journal of Physics B: Atomic, Molecular and Optical Physics* 32 (13): 3147.

Kim, Yong-Ki. 2007. 'Scaled Born cross sections for excitations of H_2 by electron impact.' *The Journal of Chemical Physics* 126 (6): 064305.

Kim, Yong-Ki, and Jean-Paul Desclaux. 2002. 'Ionization of carbon, nitrogen, and oxygen by electron impact.' *Physical Review A* 66 (1): 012708.

Kim, Y-K., W. Hwang, N. M. Weinberger, M. A. Ali, and M. Eugene Rudd. 1997. 'Electron-impact ionization cross sections of atmospheric molecules.' *The Journal of Chemical Physics* 106 (3): 1026–1033.

Kim, Yong-Ki, and Karl K. Irikura. 2000. 'Electron-impact ionization cross sections for polyatomic molecules, radicals, and ions.' *AIP Conference Proceedings* 543 (1): 220–241.

Kim, Yong-Ki, and Philip M. Stone. 2001. 'Ionization of boron, aluminium, gallium, and indium by electron impact'. *Physical Review A* 64 (5): 052707.

Kim, Yong-Ki, and M. Eugene Rudd. 1994. 'Binary-encounter-dipole model for electron-impact ionization.' *Physical Review A* 50 (5): 3954.

Kim Y-K, W. Hwang, and N. M. Weinberger. 1997. *The Journal of Chemical Physics* 106 (3): 1026–1033. See also http://physics.nist.gov/PhysRefData/Ionization

Kimura, M., O. Sueoka, C. Makochekanwa, H. Kawate, and M. Kawada. 2001. 'A comparative study of electron and positron scattering from molecules. IV. CH_3Cl, CH_3Br, and CH_3I molecules.' *The Journal of Chemical Physics* 115 (16): 7442–7449.

King, George C. 2011. 'The Use of the Magnetic Angle Changer in Atomic and Molecular Physics'. *Advances in Atomic Molecular and Optical Physics* 60: 1–64.

King, Simon J., and Stephen D. Price. 2011. 'Electron ionization of $SiCl_4$.' *The Journal of chemical physics* 134 (7): 074311.

Kleinpoppen, Hans, Bernd Lohmann, and Alexei N. Grum-Grzhimailo. 2016. *Perfect / Complete Scattering Experiments.* Berlin: Springer-Verlag.

Knudsen, H., L. Brun-Nielsen, M. Charlton, and M. R. Poulsen. 1990. 'Single ionization of H_2, He, Ne and Ar by positron impact.' *Journal of Physics B: Atomic, Molecular and Optical Physics* 23 (21): 3955.

Koch A. 1996. 'Experimentelle Untersuchungen zur elastischen Elektronenreflexion an Festkörperoberflächen und Bestimmung der mittleren freien Weglänge für inelastische Streuung im keV-Bereich', PhD thesis (in German), Physics Department Eberhard-Karls-Universitat Tubingen, Germany; as quoted in Powell, C. J. 1999. 'Consistency of calculated and measured electron inelastic mean free paths'. *Journal of Vacuum Science & Technology A: Vacuum, Surfaces, and Films* 17 (4): 1122.

Koga, Toshikatsu, Shinya Watanabe, Katsutoshi Kanayama, Ryuji Yasuda, and Ajit J. Thakkar. 1995. 'Improved Roothaan–Hartree–Fock wave functions for atoms and ions with $N \leq 54$.' *The Journal of Chemical Physics* 103 (8): 3000–3005.

Korot, Kirti, Minaxi Vinodkumar, and Harshad Bhutadia. 2012. 'Electron impact total cross sections for hydrogen molecule from 0.01 eV to 2 keV.' *Journal of Physics: Conference Series* 388 (5): 052072.

Kothari, Harshit N., and K. N. Joshipura. 2010. 'Positron scattering and ionization of neon atoms—theoretical investigations.' *Chinese Physics B* 19 (10): 103402.

Kothari, Harshit N., and K. N. Joshipura. 2012. 'Total (complete) and ionization cross sections of argon and krypton by positron impact from 15 to 2000 eV—theoretical investigations.' *Pramana J. Phys* 79.

Kothari, Harshit N., Siddharth H. Pandya, and K. N. Joshipura. 2011. 'Electron impact ionization of plasma important $SiCl_X$ (X= 1–4) molecules: theoretical cross sections.' *Journal of Physics B: Atomic, Molecular and Optical Physics* 44 (12): 125202.

Krishnakumar E. and S. K. Srivastava. 1988. 'Cross sections for electron impact ionization of O_2.' *International Journal of Mass Spectrometry and Ion Processes* 113: 1–12.

———. 1988. 'Ionisation cross sections of rare-gas atoms by electron impact.' *Journal of Physics B: Atomic, Molecular and Optical Physics* 21 (6): 1055.

———. 1990. 'Cross sections for the production of N_2^+, N^+ N_2^{2+} and N^{2+} by electron impact on N_2.' *Journal of Physics B: Atomic, Molecular and Optical Physics* 23 (11): 1893.

———. 1995. 'Ionization cross sections of silane and disilane by electron impact.' *Contributions to Plasma Physics* 35 (4-5): 395–404.

Kubo, M., D. Matsunaga, T. Suzuki, and H. Tanaka. 1981. In Proceedings of *International Conference on the Physics of Electronic and Atomic Collisions ICPEAC 12th* (Gatlinburg USA), Abstracts Ed. Datz S. Amsterdam: North-Holland. page 360.

Kwan, Ch K., Y. F. Hsieh, W. E. Kauppila, Steven J. Smith, T. S. Stein, M. N. Uddin, and M. S. Dababneh. 1983. '$e^{\pm}$– CO and $e^{\pm}$– CO_2 total cross section measurements.' *Physical Review A* 27 (3): 1328.

Kwitnewski, Stanisław, Elżbieta Ptasińska-Denga, and Czesław Szmytkowski. 2003. 'Relationship between electron-scattering grand total and ionization total cross sections.' *Radiation Physics and Chemistry* 68 1: 169–174.

Langer, Judith, Stefan Matejcik, and Eugen Illenberger. 2000. 'The nucleophilic displacement (SN_2) reaction F^- + CH_3Cl → CH_3F + Cl^- induced by resonant electron capture in gas phase clusters.' *Physical Chemistry Chemical Physics* 2 (5): 1001–1005.

Laricchia G., M. Charlton, S. Davies, C. Beling, and T. Griffith. 1987. 'The production of collimated beams of o-Ps atoms using charge exchange in positron-gas collisions'. *Journal of Physics B: Atomic Molecular and Optical Physics* 20 (3): L99.

Larsson, Mats. 2012. 'Dissociative recombination of H_3^+: 10 years in retrospect.' *Philosophical Transactions of the Royal Society* A 370: 5118–5129

Lee, Young-Sook, Young-Sil Kwak, Kyung-Chan Kim, Brian Solheim, Regina Lee, and Jaejin Lee. 2017. 'Observation of atomic oxygen O (1S) green-line emission in the summer polar upper mesosphere associated with high-energy (≥ 30 keV) electron precipitation during high-speed solar wind streams.' *Journal of Geophysical Research: Space Physics* 122 (1): 1042–1054.

Lepp, Stephen, P. C. Stancil, and A. Dalgarno. 2002. 'Atomic and molecular processes in the early Universe.' *Journal of Physics B: Atomic, Molecular and Optical Physics* 35 (10): R57.

Lesiak B., A. Jablonski, L. Zommer, A. Kosinski, G. Gergely, A. Konkol, A. Sulyok, Cs. Daroczi, and P. Nagy. 1996. 'Comparison of Al and Ni Reference Samples for Determining the Inelastic Mean Free Path of Electrons Using Different Electron Spectrometers'. *Proceedings of the European Conference on the Applications of Surface and Interface Analysis ECASIA-95*. pp 619.

Limão-Vieira, P., M. Horie, H. Kato, M. Hoshino, F. Blanco, G. García, S. J. Buckman, and H. Tanaka. 2011. 'Differential elastic electron scattering cross sections for CCl_4 by 1.5–100 eV energy electron impact'. *The Journal of Chemical Physics* 135 (23): 234309.

Limbachiya, Chetan, Minaxi Vinodkumar, Mohit Swadia, K. N. Joshipura, and Nigel Mason. 2015. 'Electron-impact total cross sections for inelastic processes for furan, tetrahydrofuran and 2, 5-dimethylfuran.' *Molecular Physics* 113 (1): 55–62.

Limbachiya, Chetan, Minaxi Vinodkumar, Mohit Swadia, and Avani Barot. 2014. 'Electron impact total cross section calculations for CH_3SH (methanethiol) from threshold to 5 keV.' *Molecular Physics* 112 (1): 101–106.

Lindsay, B. G., and M. A. Mangan. 2003. '5.1 Ionization'. In *Interactions of Photons and Electrons with Molecules. Landolt-Börnstein - Group I Elementary Particles, Nuclei and Atoms (Numerical Data and Functional Relationships in Science and Technology) Vol. 17C*, edited by Y. Itikawa. Heidelberg: Springer.

Lindsay, B. G., M. A. Mangan, H. C. Straub, and R. F. Stebbings. 2000. 'Absolute partial cross sections for electron-impact ionization of NO and NO_2 from threshold to 1000 eV.' *The Journal of Chemical Physics* 112 (21): 9404–9410.

Lindsay, B. G., R. Rejoub, and R. F. Stebbings. 2003. 'Absolute cross sections for electron-impact ionization of N_2O, H_2S, and CS_2 from threshold to 1000 eV.' *The Journal of Chemical Physics* 118 (13): 5894–5900.

Logan, Jennifer A., Michael J. Prather, Steven C. Wofsy, and Michael B. McElroy. 1981.'Tropospheric chemistry: A global perspective.' *Journal of Geophysical Research: Oceans* 86 (C8): 7210–7254.

Lotz, Wolfgang. 1970.'Electron-impact ionization cross sections for atoms up to Z= 108.' *Zeitschrift für Physik A Hadrons and nuclei* 232 (2): 101–107.

Mackus, A. J. M., A. A. Bol, and W. M. M. Kessels. 2014. `The use of atomic layer deposition in advanced nanopatterning'. *Nanoscale* 6: 10941–10960.

Maerk, T. D., and G. H. Dunn (eds). 2010. *Electron Impact Ionization*. Wien: Springer-Verlag.

Maihom, Thana, Ivan Sukuba, Ratko Janev, Kurt Becker, Tilmann Märk, Alexander Kaiser, Jumras Limtrakul, Jan Urban, Pavel Mach, and Michael Probst. 2013. 'Electron impact ionization cross sections of beryllium and beryllium hydrides.' *European Physical Journal D--Atoms, Molecules, Clusters & Optical Physics* 67 (1).

Makabe,Toshiaki, and Zoran Lj Petrovic. 2014. *Plasma electronics: applications in microelectronic device fabrication* 6. Boca Raton: CRC Press.

Mallon, P. E. 2003. 'Application to polymers.' *Principles and application of positron & positronium chemistry* 1. Singapore: World Scientific Publishing Company.

Mangan, T. P., C. G. Salzmann, J. M. C. Plane, and B. J. Murray. 2017. 'CO_2 ice structure and density under Martian atmospheric conditions.' *Icarus* 294 (201).

Margreiter, D., H. Deutsch, and T. D. Märk. 1994. 'A semiclassical approach to the calculation of electron impact ionization cross sections of atoms: from hydrogen to uranium.' *International Journal of Mass Spectrometry and Ion Processes* 139: 127–139.

Marler, Joan P., J. P. Sullivan, and Clifford M. Surko. 2005. 'Ionization and positronium formation in noble gases.' *Physical Review A* 71 (2): 022701.

Martinez, M., H. Harder, T. A. Kovacs, J. B. Simpas, J. Bassis, R. Lesher, W. H. Brune, H. Williams, C. A. Stroud, G. Frost, S. R. Hall, R. E. Shetter, B. Wert, A. Fried, B. Alicke, and J. Stutz. 2003. 'OH and HO_2 concentrations, sources, and loss rates during the Southern Oxidants Study in Nashville, Tennessee, summer 1999.' *Journal of Geophysical Research: Atmospheres* 108 (D19): 4617.

Maru, M. P., and H. S. Desai. 1975. 'A note on the effect of a short range force on the scattering of slow electrons by polar molecules.' *Journal of Physics B: Atomic and Molecular Physics* 8 (11): 1959.

Mason, N. J. 1993. 'Laser-assisted electron-atom collisions.' *Reports on Progress in Physics* 56 (10): 1275.

Mason, Nigel J., Anita Dawes, Philip D. Holtom, Robin J. Mukerji, Michael P. Davis, Bhalamurugan Sivaraman, Ralf I. Kaiser, Søren V. Hoffmann, and David A. Shaw. 2006. 'VUV spectroscopy and photo-processing of astrochemical ices: an experimental study.' *Faraday Discussions* 133: 311–329.

Mason, Nigel J., Binukumar Nair, Sohan Jeetha, and Ewelina Szmnskaa. 2014. 'Electron induced Chemistry: A new frontier in Astrochemistry'. *Faraday Discussions* 168: 235–247.

Massey, H. S. W. 1930. 'Scattering of fast electrons and nuclear magnetic moments.' *Proceedings of the Royal Society of London. Series A, Containing Papers of a Mathematical and Physical Character* 127 (806): 666–670.

May, Olivier, Juraj Fedor, and Michael Allan. 2010. 'Absolute dissociative electron attachment cross sections in acetylene and its deuterated analogue.' *CHIMIA International Journal for Chemistry* 64 (3): 173–176.

McLean, A. D., and R. S. McLean. 1981. 'Roothaan-Hartree-Fock atomic wave functions Slater basis-set expansions for Z= 55–92.' *Atomic Data and Nuclear Data Tables* 26 (3-4): 197–381.

Michaud, M., A. Wen, and L. Sanche. 2003. 'Cross sections for low-energy (1–100 eV) electron elastic and inelastic scattering in amorphous ice.' *Radiation Research* 159 (1): 3–22.

Millward, George, Steve Miller, Tom Stallard, Alan D. Aylward, and Nicholas Achilleos. 2002. 'On the dynamics of the Jovian ionosphere and thermosphere: III. The modelling of auroral conductivity.' *Icarus* 160 (1): 95–107.

Misra, Deepankar, Umesh Kadhane, Y. P. Singh, Lokesh C. Tribedi, P. D. Fainstein, and P. Richard. 2004. 'Interference effect in electron emission in heavy ion collisions with H_2 detected by comparison with the measured electron spectrum from atomic hydrogen.' *Physical Review letters* 92 (15): 153201.

Modi, Hitesh M., Manish J. Pindaria, and K. N. Joshipura. 2015. 'Scattering of Electrons/Positrons by H-atoms and H_2 Molecules under weakly Coupled Plasmas'. *Journal of Atomic, Molecular, Condensate & Nano Physics* 2 (1): 41–54.

Monks, Paul S 2005. 'Gas-phase radical chemistry in the troposphere.' *Chemical Society Reviews* 34 (5): 376–395.

Moores, D. L. 1996. 'Electron impact ionization of Be and B atoms and ions - revised data'. *Physica Scripta* T62: 19.

Mori, S., and O. Sueoka. 1994. 'Excitation and ionization cross sections of He, Ne and Ar by positron impact.' *Journal of Physics B: Atomic, Molecular and Optical Physics* 27 (18): 4349.

Mott, Nevill Francis, and Harrie Stewart Wilson Massey. 1965. *The Theory of Atomic Collisions* Vol. 35. Oxford: Clarendon Press.

Mozejko, Pawel, Grzegorz Kasperski, and Czeslaw Szmytkowski. 1996. 'Electron collisions with germane molecules. Absolute total cross section measurements from 0.75 to 250 eV.' *Journal of Physics B: Atomic, Molecular and Optical Physics* 29 (15): L571.

Możejko, P., G. Kasperski, Cz Szmytkowski, A. Zecca, G. P. Karwasz, L. Del Longo, and R. S. Brusa. 1999. 'Absolute total cross section measurements for electron scattering from silicon tetrachloride, $SiCl_4$, molecules.' *The European Physical Journal D-Atomic, Molecular, Optical and Plasma Physics* 6 (4): 481–485.

Możejko, Paweł, and Léon Sanche. 2005. 'Cross sections for electron scattering from selected components of DNA and RNA.' *Radiation Physics and Chemistry* 73 (2): 77–84.

Możejko, Paweł, Bożena Żywicka-Możejko, and Czesław Szmytkowski. 2004. 'Elastic cross section calculations for electron collisions with XY_4 (X= Si, Ge; Y= H, F, Cl, Br, I) molecules.' *Nuclear instruments and methods in physics research section b: beam interactions with materials and atoms* 196 (3-4): 245–252.

Murray, Andrew James, Martyn Hussey, Alex Knight-Percival, Sarah Jhumka, Kate L. Nixon, Matthew Harvey, and John Agomuo. 2012. 'Electron impact ionization and excitation studies of laser prepared atomic targets.' *Journal of Physics: Conference Series* (388) (1): 012009.

NIST database. http://www.nist.gov/pml/data/ionization/electron-impact-cross-sections-ionization-and-excitation-database;

NIST Chemistry WebBook, NIST Standard Reference Database Number 69. Accessed March 3, 2018. http://webbook.nist.gov/chemistry/

Naghma, Rahla, and Bobby Antony. 2013. 'Electron impact ionization cross section of C_2, C_3, Si_2, Si_3, SiC, SiC_2 and Si_2C.' *Molecular Physics* 111 (2): 269–275.

———. 2013. 'Total and elastic cross sections for methyl halides by electron impact'. *Journal of Electron Spectroscopy and Related Phenomena* 189: 17.

Naghma, Rahla, Minaxi Vinodkumar, and Bobby Antony. 2014. 'Total cross sections for O_2 and S_2 by electron impact.' *Radiation Physics and Chemistry* 97: 6–11.

Nagy, P., A. Skutlartz, and V. Schmidt. 1980. 'Absolute ionisation cross sections for electron impact in rare gases.' *Journal of Physics B: Atomic and Molecular Physics* 13 (6): 1249.

Nakano, Tohru, Hirotaka Toyoda, and Hideo Sugai. 1991. 'Electron-Impact Dissociation of Methane into CH_3 and CH_2 Radicals I. Relative Cross Sections.' *Japanese journal of applied physics* 30 (11R): 2908.

Newson, K. A., S. M. Luc, S. D. Price, and N. J. Mason. 1995. 'Electron-impact ionization of ozone'. *International Journal of Mass Spectrometry and Ion Processes* 148 (3): 203–213.

Nickel, J. C., K. Imre, D. F. Register, and S. Trajmar. 1985. 'Total electron scattering cross sections. I. He, Ne, Ar, Xe.' *Journal of Physics B: Atomic and Molecular Physics* 18 (1): 125.

Nickel, J. C., P. W. Zetner, G. Shen, and S. Trajmar. 1989. 'Principles and procedures for determining absolute differential electron-molecule (atom) scattering cross sections.' *Journal of Physics E: Scientific Instruments* 22 (9): 730.

Nishimura, H., Winifred M. Huo, M. A. Ali, and Yong-Ki Kim. 1999. 'Electron-impact total ionization cross sections of CF_4, C_2F_6, and C_3F_8.' *The Journal of Chemical Physics* 110 (8): 3811–3822.

Nishimura, Hiroyuki, Fumio Nishimura, Yoshiharu Nakamura, and Keisuke Okuda. 2003. 'Total electron scattering cross sections for simple perfluorocarbons.' *Journal of the Physical Society of Japan* 72 (5): 1080–1086.

Nishimura, H., and H. Tawara. 1994. 'Total electron impact ionization cross sections for simple hydrocarbon molecules.' *Journal of Physics B: Atomic, Molecular and Optical Physics* 27 (10): 2063.

Nogueira, J. C., I. Iga, and J. E. Chaguri. 1985. 'Total cross section measurements of electrons scattered by nitrogen and carbon dioxide in the energy range 500-3000 eV.' *Revista Brasileira de Fisica* 15 (3): 224–234.

Onthong, U., H. Deutsch, K. Becker, S. Matt, M. Probst, and T. D. Märk. 2002. 'Calculated absolute electron impact ionization cross section for the molecules CF_3X (X= H, Br, I).' *International Journal of Mass Spectrometry* 214 (1): 53–56.

Orient, O. J., I. Iga, and S. K. Srivastava. 1982. 'Elastic scattering of electrons from SO_2.' *The Journal of Chemical Physics* 77 (7): 3523–3526.

Orient, O. J., and S. K. Srivastava. 1984. 'Mass spectrometric determination of partial and total electron impact ionization cross sections of SO_2 from threshold up to 200 eV.' *The Journal of Chemical Physics* 80 (1): 140–143.

———. 1987. 'Electron impact ionisation of H_2O, CO, CO_2 and CH_4.' *Journal of Physics B: Atomic and Molecular Physics* 20 (15): 3923.

Ormonde, Stephan, Kenneth Smith, Barbara W. Torres, and Alan R. Davies. 1973. 'Configuration-interaction effects in the scattering of electrons by atoms and ions of nitrogen and oxygen.' *Physical Review A* 8 (1): 262.

Otvos, J. W., and D. P. Stevenson. 1956. 'Cross sections of molecules for ionization by electrons.' *Journal of the American Chemical Society* 78 (3): 546–551.

Pandya, Siddharth H. 2013. `Electron scattering and atomic molecular processes, theoretical studies and planetary applications'. PhD dissertation (unpublished), Sardar Patel University, Vallabh Vidyanagar, India. http://hdl.handle.net/10603/34598

Pandya, Siddharth H., and K. N. Joshipura. 2010. *PRAJÑĀ - J. Pure and App. Sci.* 18: 151.

———. Joshipura. 2014. 'Electron scattering: theory and applications'. Germany: LAP Lambert Academic Publishing.

————. 2014. 'Ionization of metastable nitrogen and oxygen atoms by electron impact: Relevance to auroral emissions.' *Journal of Geophysical Research: Space Physics* 119 (3): 2263–2268.

Pandya, Siddharth H., Foram A. Shelat, K. N. Joshipura, and Bhushit G. Vaishnav. 2012. 'Electron ionization of exotic molecular targets CN, C_2N_2, HCN, HNC and BF—Theoretical cross sections.' *International Journal of Mass Spectrometry* 323: 28–33.

Pandya, Siddharth H., B. G. Vaishnav, and K. N. Joshipura. 2012. 'Electron inelastic mean free paths in solids: A theoretical approach.' *Chinese Physics B* 21 (9): 093402.

Patel, Umang R., K. N. Joshipura, Harshit N. Kothari, and Siddharth H. Pandya. 2014. 'Electron ionization of open/closed chain isocarbonic molecules relevant in plasma processing: Theoretical cross sections.' *The Journal of Chemical Physics* 140 (4): 044302.

Penn, David R. 1976. 'Quantitative chemical analysis by ESCA.' *Journal of Electron Spectroscopy and Related Phenomena* 9 (1): 29–40.

Perdew, John P., and Alex Zunger. 1981. 'Self-interaction correction to density-functional approximations for many-electron systems.' *Physical Review B* 23 (10): 5048.

Perrin, J., and J. F. M. Aarts. 1983. 'Dissociative excitation of SiH_4, SiD_4, Si_2H_6 and GeH_4 by 0–100 eV electron impact.' *Chemical Physics* 80 (3): 351–365.

Perry, J. J., Y. H. Kim, Jane L. Fox, and H. S. Porter. 1999. 'Chemistry of the Jovian auroral ionosphere.' *Journal of Geophysical Research: Planets* 104 (E7): 16541–16565.

Politzer, Peter, Jane S. Murray, M. Edward Grice, Tore Brinck, and Shoba Ranganathan. 1991. 'Radial behavior of the average local ionization energies of atoms.' *The Journal of chemical physics* 95 (9): 6699–6704.

Poll H.U., C.Winkler, D.Margreiter, V.Grill and T.D.Maerk. 1992. 'Discrimination effects for ions with high initial kinetic energy in a Nier-type ion source and partial and total electron ionization cross sections of CF_4.' *International Journal of Mass Spectrometry and Ion Processes* 112 (1): 1–17

Posseme Nicolas. 2015. *Plasma Etching Processes for Interconnect Realization in VLSI*. London: ISTE Press.

Powel C. J. and A. Jablonski. 1999. 'Evaluation of Calculated and Measured Electron Inelastic Mean Free Paths Near Solid Surfaces.' *Journal of Physical and Chemical* Reference Data 28 (1): 19–62.

Probst, M., H. Deutsch, K. Becker, and T. D. Märkde. 2001. 'Calculations of absolute electron-impact ionization cross sections for molecules of technological relevance using the DM formalism.' *International Journal of Mass Spectrometry* 206 (1): 13–25.

Quayle, J. Rodney, and T. Ferenci. 1978. 'Evolutionary aspects of autotrophy.' *Microbiological Reviews* 42 (2): 251.

Rahman, M. A., Sumona Gangopadhyay, Chetan Limbachiya, K. N. Joshipura, and E. Krishnakumar. 2012. 'Electron ionization of NF_3.' *International Journal of Mass Spectrometry* 319: 48–54.

Raj, Deo, and Surekha Tomar. 1997. 'Electron scattering by triatomics: and OCS at intermediate energies.' *Journal of Physics B: Atomic, Molecular and Optical Physics* 30 (8): 1989.

Raju, Gorur Govinda. 2011. *Gaseous Electronics: Tables, Atoms, and Molecules*. Boca Raton: CRC Press.

————. 2016. *Dielectrics in Electric Fields*. Boca Raton: CRC press.

Ralphs, K., G. Serna, L. R. Hargreaves, M. A. Khakoo, C. Winstead, and V. McKoy. 2013. 'Excitation of the six lowest electronic transitions in water by 9–20 eV electrons.' *Journal of Physics B: Atomic, Molecular and Optical Physics* 46 (12): 125201.

Ramsauer, Carl. 1921. 'Über den Wirkungsquerschnitt der Gasmoleküle gegenüber langsamen Elektronen.' *Annalen der Physik* 369 (6): 513–540.

Rao, M. V. V. S., and S. K. Srivastava. 1992. 'Total and partial ionization cross sections of NH_3 by electron impact.' *Journal of Physics B: Atomic, Molecular and Optical Physics* 25 (9): 2175.

——— 1993. 'Electron impact ionization and attachment cross sections for H_2S.' *Journal of Geophysical Research: Planets* 98 (E7): 13137–13145.

———. 1996. 'Cross sections for the production of positive ions by electron impact on F_2.' *Journal of Physics B: Atomic, Molecular and Optical Physics* 29 (9): 1841.

Rapp, Donald, and Paula Englander-Golden. 1965. 'Total cross sections for ionization and attachment in gases by electron impact. I. Positive ionization.' *The Journal of Chemical Physics* 43 (5): 1464–1479.

Rejoub, R., B. G. Lindsay, and R. F. Stebbings. 2002. 'Electron-impact ionization of the methyl halides.' *The Journal of Chemical Physics* 117 (14): 6450–6454.

Rejoub, R., C. D. Morton, B. G. Lindsay, and R. F. Stebbings. 2003. 'Electron-impact ionization of the simple alcohols.' *The Journal of Chemical Physics* 118 (4): 1756–1760.

Rejoub, R., D. R. Sieglaff, B. G. Lindsay, and R. F. Stebbings. 2001. 'Absolute partial cross sections for electron-impact ionization of SF_6 from threshold to 1000 eV.' *Journal of Physics B: Atomic, Molecular and Optical Physics* 34 (7): 1289.

Rhodin, T. N., and J. Gadzuk. 1979. 'The nature of the surface chemical bond' New York: North-Holland; see also Penn, D. R. J. 1976. 'Quantitative chemical analysis by ESCA'. *Journal of Electron Spectroscopy and Related Phenomena* 9 (1): 29–40.

Roach, Thomas, and Anja Krieger-Liszkay. 2014. 'Regulation of photosynthetic electron transport and photoinhibition.' *Current Protein and Peptide Science* 15 (4): 351–362.

Rozum, I., P. Limao-Vieira, S. Eden, J. Tennyson, and N. J. Mason. 2006. 'Electron Interaction Cross Sections for CF_3I, C_2F_4, and CF x (x= 1–3) Radicals.' *Journal of Physical and Chemical Reference Data* 35 (1): 267–284.

Sabin, John R. 2006. *Advances in quantum chemistry: theory of the interaction of radiation with biomolecules*, 52. Cambridge: Academic Press.

Saksena, Vandana, M. S. Kushwaha, and S. P. Khare. 1997. 'Electron impact ionisation of molecules at high energies.' *International Journal of Mass Spectrometry and Ion Processes* 171 (1-3): L1–L5.

Salvat, Francesc, and Joan Parellada. 1984. 'Penetration and energy loss of fast electrons through matter.' *Journal of Physics D: Applied Physics* 17 (7): 1545.

Sanche, Leon 2005. 'Low energy electron-driven damage in biomolecules.' *The European Physical Journal D-Atomic, Molecular, Optical and Plasma Physics* 35 (2): 367–390.

Santos, J. P., and F. Parente. 'Ionisation of phosphorus, arsenic, antimony, and bismuth by electron impact.' 2008. *The European Physical Journal D-Atomic, Molecular, Optical and Plasma Physics* 47 (3): 339–350.

Schappe, R. S., P. Feng, L. W. Anderson, C. C. Lin, and T. Walker. 1995. 'Electron collision cross sections measured with the use of a magneto-optical trap.' *EPL (Europhysics Letters)* 29 (6): 439.

Schiff, L. I. 1968. *Quantum Mechanics*. New York: McGraw Hill. pp 65.

Schippers, S. 2015. 'Electron-ion merged-beam experiments at heavy-ion storage rings'. *Nuclear Instruments and Methods in Physics Research Section B: Beam Interactions with Materials and Atoms* 350: 61.

Schram, B. L., M. J. Van der Wiel, F. J. De Heer, and H. R. Moustafa. 1966. 'Absolute gross ionization cross sections for electrons (0.6–12 keV) in hydrocarbons.' *The Journal of Chemical Physics* 44 (1): 49–54.

Schulz, George J. 1963. 'Resonance in the elastic scattering of electrons in helium.' *Physical Review Letters* 10 (3): 104.

———. 1973. 'Resonances in electron impact on atoms.' *Reviews of Modern Physics* 45 (3): 378.

Schwerdtfeger, Peter. 2015. 'Table of experimental and calculated static dipole polarizabilities for the electronic ground states of the neutral elements (in atomic units)' Accessed 31 August 2018, http://ctcp.massey.ac.nz/dipole-polarizabilities

Shah, M. B., D. S. Elliott, and H. B. Gilbody. 1987. 'Pulsed crossed-beam study of the ionisation of atomic hydrogen by electron impact.' *Journal of Physics B: Atomic and Molecular Physics* 20 (14): 3501.

Shah, M. B., D. S. Elliott, P. McCallion, and H. B. Gilbody. 1988. 'Single and double ionisation of helium by electron impact.' *Journal of Physics B: Atomic, Molecular and Optical Physics* 21 (15): 2751.

Shelat, F. A., K. N. Joshipura, K. L. Baluja, P. Bhowmik, and H. N. Kothari. 2011. 'Electron scattering with LiH molecule including a realistic dipole potential.' *Indian Journal of Physics* 85 (12): 1739–1748.

Shi, D. H., J. F. Sun, Z. L. Zhu, and Y. F. Liu. 2010. 'Total cross sections of electron scattering by molecules NF_3, PF_3, $N(CH_3)_3$, $P(CH_3)_3$, $NH(CH_3)_2$, $PH(CH_3)_2$, NH_2CH_3 and PH_2CH_3 at 30–5000 eV.' *The European Physical Journal D-Atomic, Molecular, Optical and Plasma Physics* 57 (2): 179–186.

Shipman, M., S. Brawley, L. Sarkadi, and G. Laricchia. 2017. 'Resonant scattering of positronium as a quasifree electron'. *Physical Review A* 95 (3): 032704.

Shyn, T. W., and G. R. Carignan. 1980. 'Angular distribution of electrons elastically scattered from gases: 1.5-400 eV on N_2. II.' *Physical Review A* 22 (3): 923.

Sieradzka, Agnieszka, and Jimena D. Gorfinkiel. 2017.'Theoretical study of resonance formation in microhydrated molecules. I. Pyridine-$(H_2O)_n$, n= 1, 2, 3, 5.' *The Journal of Chemical Physics* 147 (3): 034302.

———. 2017. 'Theoretical study of resonance formation in microhydrated molecules.' II. Thymine-(H_2O) n, n= 1, 2, 3, 5.' *The Journal of Chemical Physics* 147 (3): 034303.

Šiller, L., M. T. Sieger, and T. M. Orlando. 2003. 'Electron-stimulated desorption of D_2O coadsorbed with CO_2 ice at VUV and EUV energies.' *The Journal of Chemical Physics* 118 (19): 8898–8904.

Singh, Jasmeet, and K. L. Baluja. 2014. 'Electron-impact study of the O_2 molecule using the R-matrix method.' *Physical Review A* 90 (2): 022714.

Singh, Suvam, and Bobby Antony. 2017. 'Study of inelastic channels by positron impact on simple molecules.' *Journal of Applied Physics* 121 (24): 244903.

Singh, Suvam, Sangita Dutta, Rahla Naghma, and Bobby Antony. 2016. 'Theoretical Formalism To Estimate the Positron Scattering Cross Section.' *The Journal of Physical Chemistry A* 120 (28): 5685–5692.

———. 2017. 'Positron scattering from simple molecules'. *Journal of Physics B: Atomic Molecular and Optical Physics* 50 (13): 135202.

Singh, Suvam, Rahla Naghma, Jaspreet Kaur, and Bobby Antony. 2016. 'Calculation of total and ionization cross sections for electron scattering by primary benzene compounds.' *The Journal of Chemical Physics* 145 (3): 034309.

Sivaraman, B., A. M. Mebel, N. J. Mason, D. Babikov, and R. I. Kaiser. 2011. 'On the electron-induced isotope fractionation in low temperature $^{32}O_2/^{36}O_2$ ices—ozone as a case study.' *Physical Chemistry Chemical Physics* 13 (2): 421–427.

Sivaraman, B., S. Ptasinska, S. Jheeta, and N. J. Mason. 2008. 'Electron irradiation of solid nitrous oxide.' *Chemical Physics Letters* 460 (1): 108–111.

Smith, A. C. H., E. Caplinger, R. H. Neynaber, Erhard W. Rothe, and S. M. Trujillo. 1962. 'Electron impact ionization of atomic nitrogen.' *Physical Review* 127 (5): 1647.

Smith, O. I. 1983. 'Cross sections for formation of parent and fragment ions by electron impact from C_2N_2.' *International Journal of Mass Spectrometry and Ion processes* 54 (1-2): 55

Song, Mi-Young, Jung-Sik Yoon, Hyuck Cho, Yukikazu Itikawa, Grzegorz P. Karwasz, Viatcheslav Kokoouline, Yoshiharu Nakamura, and Jonathan Tennyson. 2015. 'Cross sections for electron collisions with methane.' *Journal of Physical and Chemical Reference Data* 44 (2): 023101.

Song, Mi-Young, Jung-Sik Yoon, Hyuck Cho, Grzegorz P. Karwasz, Viatcheslav Kokoouline, Yoshiharu Nakamura, and Jonathan Tennyson. 2017. 'Cross Sections for Electron Collisions with Acetylene.' *Journal of Physical and Chemical Reference Data* 46 (1): 013106.

Sorokin, A. A., L. A. Shmaenok, S. V. Bobashev, B. Möbus, M. Richter, and G. Ulm. 2000. 'Measurements of electron-impact ionization cross sections of argon, krypton, and xenon by comparison with photoionization.' *Physical Review A* 61 (2): 022723.

Sorokin, A. A., L. A. Shmaenok, S. V. Bobashev, B. Möbus, and G. Ulm. 1998. 'Measurements of electron-impact ionization cross sections of neon by comparison with photoionization.' *Physical Review A* 58 (4): 2900.

Srivastava, Rajesh, R. K. Gangwar, and A. D. Stauffer. 2009. 'Excitation of the $6^{1,3}P_1$ states of mercury by spin-resolved electron impact.' *Physical Review A* 80 (2): 022718.

Staszewska, G., D. W. Schwenke, and D. G. Truhlar. 1984. 'Complex optical potential model for electron–molecule scattering, elastic scattering, and rotational excitation of H_2 at 10–100 eV'. *Journal of Chemical Physics* 81 (1): 335; see also Blanco, F. and G. Garcia. 2002. 'Improvements on the imaginary part of a non-empirical model potential for electron scattering (30 to 10000 eV energies)'. *Physics Letters A* 295 (4): 178–184.

———. 1984. 'Investigation of the shape of the imaginary part of the optical-model potential for electron scattering by rare gases.' *Physical Review A* 29 (6): 3078.

Stephen, K., and T. D. Märk. 1984. 'Absolute partial electron impact ionization cross sections of Xe from threshold up to 180 eV.' *The Journal of Chemical Physics* 81 (7): 3116–3117.

Stevie, F. A., and M. J. Vasile. 1981. 'Electron impact ionization cross sections of F_2 and Cl_2.' *The Journal of Chemical Physics* 74 (9): 5106–5110.

Stingl, E. 1972. 'The ionization of boron and the isoelectronic ions carbon II, nitrogen III, and oxygen IV by electron impact'. *Journal of Physics B: Atomic and Molecular Physics* 5 (6): 1160.

Straub, H. C., B. G. Lindsay, K. A. Smith, and R. F. Stebbings. 1996. 'Absolute partial cross sections for electron-impact ionization of CO_2 from threshold to 1000 eV'. *Journal of Chemical Physics* 105 (10): 4015.

———. 1998. 'Absolute partial cross sections for electron-impact ionization of H_2O and D_2O from threshold to 1000 eV.' *The Journal of Chemical Physics* 108 (1): 109–116.

Straub, H. C., P. Renault, B. G. Lindsay, K. A. Smith, and R. F. Stebbings. 1996. 'Absolute partial cross sections for electron-impact ionization of H_2, N_2, and O_2 from threshold to 1000 eV.' *Physical Review A* 54 (3): 2146.

Sueoka, O., and S. Mori. 1986. 'Total cross sections for low and intermediate energy positrons and electrons colliding with CH_4, C_2H_4 and C_2H_6 molecules.' *Journal of Physics B: Atomic and Molecular Physics* 19 (23): 4035.

————. 1989. 'Total cross section measurements for 1-400 eV positrons and electrons in C_2H_2.' *Journal of Physics B: Atomic, Molecular and Optical Physics* 22 (6): 963.

Sueoka, O., S. Mori, and A. Hamada. 1994. 'Total cross section measurements for positrons and electrons colliding with molecules. I. SiH_4 and CF_4.' *Journal of Physics B: Atomic, Molecular and Optical Physics* 27 (7): 1453.

Sunshine, Gabriel, Bertrand B. Aubrey, and Benjamin Bederson. 1967. 'Absolute measurements of total cross sections for the scattering of low-energy electrons by atomic and molecular oxygen.' *Physical Review* 154 (1): 1.

Swadia, Mohit S. 2017. Private communication. (unpublished).

Szmytkowski, Cz., and E. P. Denga. 2001. 'Fluorination effects in electron-scattering processes - comparative study'. *Vacuum* 63 (4): 545.

Szmytkowski, Czesław, Alicja Domaracka, Paweł Możejko, Elżbieta Ptasińska-Denga, Łukasz Kłosowski, Michał Piotrowicz, and Grzegorz Kasperski. 2004. 'Electron collisions with nitrogen trifluoride (NF_3) molecules.' *Physical Review A* 70 (3): 032707.

Szmytkowski, Czesiaw, and A. M. Krzysztofowicz. 1995. 'Electron scattering from isoelectronic, N_e= 18, CH_3X molecules (X= F, OH, NH_2 and CH_3).' *Journal of Physics B: Atomic, Molecular and Optical Physics* 28 (19): 4291.

Szmytkowski, Czesław, and Stanisław Kwitnewski. 2002. 'Total cross sections for electron scattering with some C_3 hydrocarbons.' *Journal of Physics B: Atomic, Molecular and Optical Physics* 35 (17): 3781.

Szmytkowski, Czesław, and Krzysztof Maciag. 1986. 'Absolute total electron-scattering cross section of SO_2.' *Chemical Physics Letters* 124 (5): 463–466.

————. 1991. 'Total cross section for electron impact on nitrogen monoxide.' *Journal of Physics B: Atomic, Molecular and Optical Physics* 24 (19): 4273.

Szmytkowski, Czeslaw, Krzysztof Maciag, and Grzegorz Karwasz. 1996. 'Absolute electron-scattering total cross section measurements for noble gas atoms and diatomic molecules.' *Physica Scripta* 54 (3): 271.

Szmytkowski, Czesław, and Paweł Możejko. 2006. 'Electron-scattering total cross sections for triatomic molecules: NO_2 and H_2O.' *Optica Applicata* 36 (4).

Szmytkowski, Czeslaw, Pawel Mozejko, and Grzegorz Kasperski. 1997. 'Low-and intermediate-energy total electron scattering cross sections for SiH_4 and $GeCl_4$ molecules.' molecules.' *Journal of Physics B: Atomic, Molecular and Optical Physics* 30 (19): 4363.

————. 1998. 'Experimental absolute total cross sections for low-energy electron collisions with tetrahedral compounds of germanium.' *Journal of Physics B: Atomic, Molecular and Optical Physics* 31 (17): 3917.

Szmytkowski, Czesław, Paweł Możejko, Stanisław Kwitnewski, Alicja Domaracka, and Elżbieta Ptasińska-Denga. 2006. 'Electron collision with sulfuryl chloride (SO_2Cl_2) molecule.' *Journal of Physics B: Atomic, Molecular and Optical Physics* 39 (11): 2571.

Szmytkowski, Cz., P. Mozejko, S. Kwitnewski, E. Ptasinska-Denga, and A. Domaracka. 2005. 'Cross sections for electron scattering from sulfuryl chloride fluoride (SO_2ClF) molecules'. *Journal of Physics B: Atomic Molecular and Optical Physics* 38 (16): 2945.

Szmytkowski, Czesław, Paweł Możejko, Elżbieta Ptasińska-Denga, and Agnieszka Sabisz. 2010. 'Cross sections for electron scattering from furan molecules: Measurements and calculations.' *Physical Review A* 82 (3): 032701.

Szmytkowski, Czesław, Paweł Możejko, Mateusz Zawadzki, and Elżbieta Ptasińska-Denga. 2014. 'Scattering of electrons by a 1, 2-butadiene (C_4H_6) molecule: measurements and calculations.' *Journal of Physics B: Atomic, Molecular and Optical Physics* 48 (2): 025201.

Szymańska, Ewelina, Nigel J. Mason, E. Krishnakumar, Carolina Matias, Andreas Mauracher, Paul Scheier, and Stephan Denifl. 2014. 'Dissociative electron attachment and dipolar dissociation in ethylene.' *International Journal of Mass Spectrometry* 365: 356–364.

Tanaka, Hiroshi, Yoshio Tachibana, Masashi Kitajima, Osamu Sueoka, Hideki Takaki, Akira Hamada, and Mineo Kimura. 1999. 'Total cross sections of electron and positron collisions with C_3F_8 and C_3H_8 molecules and differential elastic and vibrational excitation cross sections by electron impact on these molecules.' *Physical Review A* 59 (3): 2006.

Tanuma, Shigeo, Cedric J. Powell, and David R. Penn. 1988. 'Calculations of electron inelastic mean free paths for 31 materials.' *Surface and Interface Analysis* 11 (11): 577–589.

———. 2011. 'Calculations of electron inelastic mean free paths. IX. Data for 41 elemental solids over the 50 eV to 30 keV range.' *Surface and Interface Analysis* 43 (3): 689–713.

Tarnovsky, V., H. Deutsch, and K. Becker. 1997. 'Cross sections for the electron impact ionization of NDx (x= 1– 3).' *International Journal of Mass Spectrometry and Ion processes* 167: 69–78.

———. 1998. 'Electron impact ionization of the hydroxyl radical.' *The Journal of Chemical Physics* 109 (3): 932–936.

Tarnovsky, V., A. Levin, and K. Becker. 1994. 'Absolute cross sections for the electron impact ionization of the NF2 and NF free radicals.' *The Journal of Chemical Physics* 100 (8): 5626–5630.

Tarnovsky, V., A. Levin, Kurt Becker, R. Basner, and M. Schmidt. 1994. 'Electron impact ionization of the NF_3 molecule.' *International Journal of Mass Spectrometry and Ion Processes* 133 (2-3): 175–185.

Tarnovsky, V., A. Levin, H. Deutsch, and K. Becker. 1995. 'Electron-impact ionization of the SO free radical.' *The Journal of Chemical Physics* 102 (2): 770–773.

———. 1996. 'Electron impact ionization of CD_x (x= 1-4).' *Journal of Physics B: Atomic, Molecular and Optical Physics* 29 (1): 139.

Tawara, Hiro, Y. Itikawa, H. Nishimura, and Masatoshi Yoshino. 1990. 'Cross sections and related data for electron collisions with hydrogen molecules and molecular ions.' *Journal of Physical and Chemical Reference Data* 19 (3): 617–636.

Tennyson, Jonathan. 2010. 'Electron–molecule collision calculations using the R-matrix method.' *Physics Reports* 491 (2): 29–76.

Tennyson, Jonathan, Daniel B. Brown, James J. Munro, Iryna Rozum, Hemal N. Varambhia, and Natalia Vinci. 2007. 'Quantemol-N: an expert system for performing electron molecule collision calculations using the R-matrix method.' *Journal of Physics: Conference Series*, 86 (1) 012001.

Tennyson, Jonathan, and C. S. Trevisan. 2002. 'Low energy electron collisions with molecular hydrogen.' *Contributions to Plasma Physics* 42 (6-7): 573–577.

Thompson, W. R., M. B. Shah, and H. B. Gilbody. 1995. 'Single and double ionization of atomic oxygen by electron impact.' *Journal of Physics B: Atomic, Molecular and Optical Physics* 28 (7): 1321.

Thorn, Penny Anne, Michael James Brunger, P. J. O. Teubner, N. Diakomichalis, T. Maddern, M. A. Bolorizadeh, W. R. Newell et al. 2007. 'Cross sections and oscillator strengths for electron-impact excitation of the A^1B_1 electronic state of water.' *The Journal of Chemical Physics* 126 (6): 064306.

Toyoda, Hirotaka, Makoto Iio, and Hideo Sugai. 1997. 'Cross Section Measurements for Electron-Impact Dissociation of C_4F_8 into Neutral and Ionic Radicals.' *Japanese journal of Applied Physics* 36 6R: 3730.

Trajmar, S., David F. Register, and Ara Chutjian. 1983. Electron scattering by molecules II. Experimental methods and data. *Physics Reports* 97 (5): 219–356.

Tsai, J. S., L. Lebow, and D. A. L. Paul. 1976. 'Measurement of total cross sections (e⁺, Ne) and (e⁺, Ar).' *Canadian Journal of Physics* 54 (17): 1741–1748.

Tsujii, H. 2017. 'Overview of Carbon-ion Radiotherapy'. *Journal of Physics: Conference Series* 777 (1): 012032.

Tsujii, Hirohiko, and Tadashi Kamada. 2012. 'A review of update clinical results of carbon ion radiotherapy.' *Japanese Journal of Clinical Oncology* 42 (8): 670–685.

Tung, C. J., J. C. Ashley, and R. H. Ritchie. 1979. 'Electron inelastic mean free paths and energy losses in solids II: Electron gas statistical model.' *Surface Science* 81 (2): 427–439.

Turner, J. E., H. G. Paretzke, R. N. Hamm, H. A. Wright, and R. H. Ritchie. 1982. 'Comparative study of electron energy deposition and yields in water in the liquid and vapor phases.' *Radiation Research* 92 (1): 47–60.

Tîmneanu, Nicuşor, Carl Caleman, Janos Hajdu, and David van der Spoel. 2004. 'Auger electron cascades in water and ice.' *Chemical Physics* 299 (2): 277–283.

Utke, Ivo, Patrik Hoffmann, and John Melngailis. 2008. 'Gas-assisted focused electron beam and ion beam processing and fabrication.' *Journal of Vacuum Science & Technology B: Microelectronics and Nanometer Structures Processing, Measurement, and Phenomena* 26 (4): 1197–1276.

Vacher, J. R., F. Jorand, N. Blin-Simiand, and S. Pasquiers. 2009. 'Electron impact ionization of formaldehyde.' *Chemical Physics Letters* 476 (4): 178–181.

Vallance, Claire, Sean A. Harris, James E. Hudson, and Peter W. Harland. 1997. 'Absolute electron impact ionization cross sections for CH_3X, where X= H, F, Cl, Br, and I.' *Journal of Physics B: Atomic, Molecular and Optical Physics* 30 (10): 2465.

Verma, Pankaj, Jaspreet Kaur, and Bobby Antony. 2017. 'Electron-silane scattering cross section for plasma assisted processes.' *Physics of Plasmas* 24 (3): 033501.

Vinodkumar, Minaxi, Avani Barot, and Bobby Antony. 2012. 'Electron impact total cross section for acetylene over an extensive range of impact energies (1 eV–5000 eV).' *The Journal of Chemical Physics* 136 (18): 184308.

Vinodkumar, Minaxi, Harshad Bhutadia, Chetan Limbachiya, and K. N. Joshipura. 2011. 'Electron impact total ionization cross sections for H_2S, PH_3, HCHO and HCOOH.' *International Journal of Mass Spectrometry* 308 (1): 35–40.

Vinodkumar, Minaxi, Rucha Dave, Harshad Bhutadia, and Bobby K. Antony. 2010. 'Electron impact total ionization cross sections for halogens and their hydrides'. *International Journal of Mass Spectrometry* 292 (1): 7–13.

Vinodkumar, Minaxi, K. N. Joshipura, Chetan Limbachiya, and Bobby Antony. 2006. 'Electron impact total and ionization cross sections for some hydrocarbon molecules and radicals.' *The European Physical Journal D-Atomic, Molecular, Optical and Plasma Physics* 37 (1): 67–74.

Vinodkumar, Minaxi, K. N. Joshipura, Chetan Limbachiya, and Nigel Mason. 2006. 'Theoretical calculations of the total and ionization cross sections for electron impact on some simple biomolecules.' *Physical Review A* 74 (2): 022721.

Vinodkumar, Minaxi, K. N. Joshipura, and Nigel Mason. 2006. 'Modelling electron interactions: a semi-rigorous method.' *Acta Physica Slovaca* 56: 521–529.

Vinodkumar, Minaxi, Kirti Korot, and P. C. Vinodkumar. 2010. 'Complex scattering potential–ionization contribution (CSP-ic) method for calculating total ionization cross sections on electron impact.' *The European Physical Journal D-Atomic, Molecular, Optical and Plasma Physics* 59 (3): 379–387.

————. 2011. 'Computation of the electron impact total ionization cross sections of $C_nH_{(2n+1)}$ OH molecules from the threshold to 2keV energy range.' *International Journal of Mass Spectrometry* 305 (1): 26–29.

Vinodkumar, Minaxi, Chetan Limbachiya, Bobby Antony, and K. N. Joshipura. 2007. 'Calculations of elastic, ionization and total cross sections for inert gases upon electron impact: threshold to 2 keV.' *Journal of Physics B: Atomic, Molecular and Optical Physics* 40 (16): 3259.

Vinodkumar, Minaxi, Chetan Limbachiya, and Harshad Bhutadia. 2010. 'Electron impact calculations of total ionization cross sections for environmentally sensitive diatomic and triatomic molecules from threshold to 5 keV'. *Journal of Physics B: Atomic Molecular and Optical Physics* 43 (1): 015203.

Vinodkumar, Minaxi, Chetan Limbachiya, K. N. Joshipura, and Nigel Mason. 2011. 'Electron impact calculations of total elastic cross sections over a wide energy range–0.01 eV to 2 keV for CH_4, SiH_4 and H_2O.' *The European Physical Journal D-Atomic, Molecular, Optical and Plasma Physics* 61 (3): 579–585.

Vinodkumar, Minaxi, Chetan Limbachiya, Kirti Korot, and K. N. Joshipura. 2008. 'Theoretical electron impact elastic, ionization and total cross sections for silicon hydrides, SiH_x (x= 1, 2, 3, 4) and disilane, Si_2H_6 from threshold to 5 keV.' *The European Physical Journal D-Atomic, Molecular, Optical and Plasma Physics* 48 (3): 333–342.

Vinodkumar, Minaxi, Chetan Limbachiya, Kirti Korot, K. N. Joshipura, and Nigel Mason. 2008. 'Electron impact calculations of total and ionization cross sections for Germanium Hydrides $(GeH_X; X= 1–4)$ and Digermane, Ge_2H_6.' *International Journal of Mass Spectrometry* 273 (3): 145–150.

Vuitton, Véronique, P. Lavvas, R. V. Yelle, M. Galand, A. Wellbrock, G. R. Lewis, A. J. Coates, and J-E. Wahlund. 2009. 'Negative ion chemistry in Titan's upper atmosphere.' *Planetary and Space Science* 57 (13): 1558–1572.

Waldrop, M. Mitchell. 2016. 'More than moore.' *Nature* 530 (7589): 144–148.

Weisstein, Eric W. *Circle-Circle Intersection*; MathWorld-A Wolfram Web Resource. Accessed 31 August 2018. http://mathworld.wolfram.com/Circle-CircleIntersection.html

Wetzel, Robert C., Frank A. Baiocchi, Todd R. Hayes, and Robert S. Freund. 1987. 'Absolute cross sections for electron-impact ionization of the rare-gas atoms by the fast-neutral-beam method.' *Physical Review A* 35 (2): 559.

Winters, H. F. 1966. 'Ionic adsorption and dissociation cross section for nitrogen.' *The Journal of Chemical Physics* 44 (4): 1472–1476.

Yadav, Hitesh, Minaxi Vinodkumar, Chetan Limbachiya, and P. C. Vinodkumar. 2017. 'Scattering of electrons with formyl radical.' *Molecular Physics* 115 (8): 952–961.

Yoon, Jung-Sik, Mi-Young Song, Jeong-Min Han, Sung Ha Hwang, Won-Seok Chang, Bong Ju Lee, and Yukikazu Itikawa. 2008. 'Cross sections for electron collisions with hydrogen molecules.' *Journal of Physical and Chemical Reference Data* 37 (2): 913–931.

Zakrzewski, V. G., and J. V. Ortiz. 1996. 'Dichlorobenzene Ionization Energies.' *The Journal of Physical Chemistry* 100 (33): 13979–13984.

Zammit, Mark C., Jeremy S. Savage, Dmitry V. Fursa, and Igor Bray. 2016. 'Complete solution of electronic excitation and ionization in electron-hydrogen molecule scattering.' *Physical review letters* 116 (23): 233201.

Zecca, Antonio, Grzegorz P. Karwasz, and Roberto S. Brusa. 1992. 'Total-cross section measurements for electron scattering by NH_3, SiH_4, and H_2S in the intermediate-energy range.' *Physical Review A* 45(5): 2777.

————. 1996. 'One century of experiments on electron-atom and molecule scattering: A critical review of integral cross sections.' *La Rivista del Nuovo Cimento (1978-1999)* 19 (3): 1–146.

Zecca, Antonio, Grzegorz Karwasz, Roberto S. Brusa, and Czeslaw Szmytkowski. 1991. 'Absolute total cross sections for electron scattering on CH_4 molecules in the 1-4000 eV energy range.' *Journal of Physics B: Atomic, Molecular and Optical Physics* 24 (11): 2747.

Zecca, Antonio, Jose C. Nogueira, Grzegorz P. Karwasz, and R. S. Brusa. 1995. 'Total cross sections for electron scattering on NO_2, OCS, SO_2 at intermediate energies.' *Journal of Physics B: Atomic, Molecular and Optical Physics* 28 (3): 477.

Zhang, Xianzhou, Jinfeng Sun, and Yufang Liu. 1992. 'A new approach to the correlation polarization potential-low-energy electron elastic scattering by He atoms.' *Journal of Physics B: Atomic, Molecular and Optical Physics* 25 (8): 1893.

Zheng, Lianjun, Nicholas F. Polizzi, Adarsh R. Dave, Agostino Migliore, and David N. Beratan. 2016. 'Where is the electronic oscillator strength? Mapping oscillator strength across molecular absorption spectra.' *The Journal of Physical Chemistry A* 120 (11): 1933–1943.

Zhou, S., H. Li, W. E. Kauppila, C. K. Kwan, and T. S. Stein. 1997. 'Measurements of total and positronium formation cross sections for positrons and electrons scattered by hydrogen atoms and molecules.' *Physical Review A* 55 (1): 361.

Ziaja, Beata, Richard A. London, and Janos Hajdu. 2006. 'Ionization by impact electrons in solids: Electron mean free path fitted over a wide energy range.' *Journal of Applied Physics* 99 (3): 033514.

INDEX